American Book Company
The Standards Experts

PASSING THE ALABAMA HIGH SCHOOL GRADUATION EXAM IN MATHEMATICS

2009 Revision

ERICA DAY

COLLEEN PINTOZZI

AMERICAN BOOK COMPANY

P. O. BOX 2638

WOODSTOCK, GEORGIA 30188-1383

TOLL FREE 1 (888) 264-5877   PHONE (770) 928-2834

TOLL FREE FAX 1 (866) 827-3240

WEB SITE: www.americanbookcompany.com

Acknowledgements

In preparing this book, we would like to acknowledge Mary Stoddard and Eric Field for their contributions in developing graphics for this book and Camille Woodhouse for her contributions in editing this book. We would also like to thank our many students whose needs and questions inspired us to write this text.

Printed in the United States of America
03/09   06/07   08/06   09/05   10/04   06/03

Contents

Contents

Contents

Preface

Passing the NEW Alabama High School Graduation Exam in Mathematics will help you review and learn important concepts and skills related to high school mathematics. To help identify which areas are of greater challenge for you, first take the diagnostic test, then complete the evaluation chart with your instructor in order to help you identify the chapters which require your careful attention. When you have finished your review of all of the material your teacher assigns, take the practice tests to evaluate your understanding of the material presented in this book. **The materials in this book are based on the mathematics standards published by the Alabama Department of Education. The complete list of standards is located in the answer key. Each question in the Diagnostic and Practice Tests is referenced to a standard, as is the beginning of each chapter.**

This book contains several sections. These sections are as follows: 1) A Diagnostic Test; 2) Chapters that teach the concepts and skills for *Passing the NEW Alabama High School Graduation Exam in Mathematics*; 3) Two Practice Tests. Answers to the tests and exercises are in a separate manual.

ABOUT THE AUTHORS

Erica Day has a Bachelor of Science Degree in Mathematics and is working on a Master of Science Degree in Mathematics. She graduated with high honors from Kennesaw State University in Kennesaw, Georgia. She has also tutored all levels of mathematics, ranging from high school algebra and geometry to university-level statistics, calculus, and linear algebra. She is currently writing and editing mathematics books for American Book Company, where she has coauthored numerous books, such as *Passing the Georgia Algebra I End of Course*, *Passing the Georgia High School Graduation Test in Mathematics*, *Passing the Arizona AIMS in Mathematics*, and *Passing the New Jersey HSPA in Mathematics*, to help students pass graduation and end of course exams.

Colleen Pintozzi has taught mathematics at the middle school, junior high, senior high, and adult level for 22 years. She holds a B.S. degree from Wright State University in Dayton, Ohio and has done graduate work at Wright State University, Duke University, and the University of North Carolina at Chapel Hill. She is the author of many mathematics books including such best-sellers as *Basics Made Easy: Mathematics Review*, *Passing the New Alabama Graduation Exam in Mathematics*, *Passing the Louisiana LEAP 21 GEE*, *Passing the Indiana ISTEP+ GQE in Mathematics*, *Passing the Minnesota Basic Standards Test in Mathematics*, and *Passing the Nevada High School Proficiency Exam in Mathematics*.

• R E F E R E N C E   P A G E •

Use the information below to answer question on the Alabama High School Graduation Exam.

Some Abbreviations Used in Formulas

b_1, b_2 = bases of a trapezoid
  b = base of a polygon
  h = height or altitude
  l = length
  w = width

⌐ = symbol for a right angle

$m\angle$ = the measure of an angle

A = area
C = circumference
r = radius
d = diameter
π = 3.14
P = perimeter
D = distance
M = midpoint
m = slope

S.A. = surface area
V = volume
B = area of the base
S = sum of interior angles of a complex polygon
n = number of sides of a convex polygon

Formulas

Triangle: $A = \frac{1}{2}bh$

Parallelogram: $A = bh$

Rectangle: $A = lw$

Trapezoid: $A = \frac{1}{2}h(b_1 + b_2)$

Circle: $C = \pi d$
          $C = 2\pi r$
          $A = \pi r^2$

Distance = rate · time

Interest = principal · rate · time

Distance Formula: $D = \sqrt{(x_2 - x_1)^2 + (y_2 - y_1)^2}$

Midpoint Formula: $M = \left(\dfrac{x_1 + x_2}{2}, \dfrac{y_1 + y_2}{2} \right)$

Slope Formula: $m = \dfrac{y_2 - y_1}{x_2 - x_1}$

Sum of Measures of Interior Angles of a Convex Polygon: $S = 180(n - 2)$

Quadratic Formula: $x = \dfrac{-b \pm \sqrt{b^2 - 4ac}}{2a}$

Pythagorean Theorem: $c^2 = a^2 + b^2$

	Surface Area	Volume
Rectangular Prism	S.A. = Ph + 2B or S.A. = 2(wh + lh + lw)	V = Bh or V = lwh
Cylinder	S.A. = $2\pi rh + 2\pi r^2$	V = $\pi r^2 h$

Forms of Equations

Standard form of an equation of a line: $Ax + By = C$
Slope-intercept form of an equation of a line: $y = mx + b$
Point-slope form of an equation of a line: $y - y_1 = m(x - x_1)$

Diagnostic Test

1. Simplify: $12 + (5 \cdot 2)^2 \cdot 14$

   A 152
   B 1,412
   C 1,568
   D 16,184

        I-1

2. Simplify: $\dfrac{(-106) + 20 \cdot 3 + 15 \cdot 2}{8} =$

   A -2
   B 2
   C 16
   D -16

        I-1

3. Simplify: $\dfrac{75 \cdot 4 - 30 \cdot 5}{2 \cdot 5}$

   A 1.5
   B 15
   C 150
   D 1500

        I-1

4. Add $-6y^2 + 4y - 3$ and $4y^2 - 2y + 5$.

   A $-2y^2 + 2y + 2$
   B $-2y^2 - 2y - 2$
   C $-2y^2 + 2y - 2$
   D $-2y^2 - 2y + 2$

        I-1

5. Simplify: $\dfrac{4x - 1}{3} - \dfrac{2x - 8}{6}$

   A $2x + 2$
   B $2x + 1$
   C $x + 1$
   D $x + 2$

        I-2

6. Simplify $3\left(-x^2 + 4x + 1\right) - 2\left(3x^2 + 2x - 2\right)$.

   A $-3x^2 - 16x + 1$
   B $-6x^2 + 12x - 4$
   C $-9x^2 + 8x + 7$
   D $-12x^2 + 16x - 3$

        I-2

7. Isabella is simplifying this expression:

$$2(5a + 3b - c) - 5(4a - 2b - 3c)$$

The expression above is equivalent to which of the following expressions?

   A $-10a + 16b + 13c$
   B $-10a - 4b - 4c$
   C $30a + b + 2c$
   D $30a - 4b - 17c$

        I-2

8. Simplify: $(3x^2 - 5x + 6) - (x^2 + 4x - 7)$

   A $4x^2 - x - 1$
   B $4x^2 - x + 13$
   C $2x^2 - 9x - 1$
   D $2x^2 - 9x + 13$

        I-2

9. Multiply the polynomials $(6x + 2)(x - 1)$.

   A $6x^2 - 4x - 2$
   B $6x^2 - 8x + 3$
   C $2x^2 + 3x - 2$
   D $12x^2 + 8x + 4$

        I-3

10. Simplify: $4x^3 (4x)^2$

   A $16x^5$
   B $16x^6$
   C $64x^5$
   D $64x^6$

        I-3

11. Multiply and simplify:
$(3x + 2)(x - 4)$

A $3x^2 - 10x - 8$
B $3x^2 + 5x - 8$
C $3x^2 + 5x - 6$
D $8x^2 - 2$

I-3

12. Simplify: $(2x)(3y) - (-x)(5y)$

A xy
B $11xy$
C $30x^2y^2$
D $3x - 2y$

I-3

13. What term can be factored out of $24x^3y^2 - 18x^2y^2 + 12xy^6$?

A $6x^2y^2$
B $6xy^2$
C x^3y
D xy^2

I-4

14. Factor: $2x^2 - 2x - 112$

A $2(x + 7)(x - 8)$
B $(2x - 7)(2x - 8)$
C $(2x - 5)(x + 10)$
D $3(x + 3)(x - 2)$

I-4

15. Find the solution to $4m^2 = 9m + 9$.

A $\left\{-\frac{3}{2}, \frac{3}{2}\right\}$
B $\left\{3, -\frac{3}{4}\right\}$
C $\left\{-3, \frac{3}{4}\right\}$
D $\left\{-1, \frac{1}{4}\right\}$

II-2

16. Factor: $9x^2 + 15x - 14$

A $(9x - 1)(x + 7)$
B $(3x + 2)(3x - 7)$
C $(3x - 2)(3x - 7)$
D $(3x - 2)(3x + 7)$

I-4

17. Factor the polynomial $x^3 - 4x^2 - 21x$.

A $(x + 3)(x - 7)$
B $x(x + 3)(x - 7)$
C $(x + 3)(x^2 - 7)$
D $(x^2 + 3)(x - 7)$

I-4

18. Solve: $\dfrac{2x - 1}{3} = \dfrac{x + 4}{6}$

A $x = 6$
B $x = 3$
C $x = 2$
D $x = 1$

II-1

19. Solve: $6 - 2(5y - 1) = 18$

A $y = 2$
B $y = -2$
C $y = 1$
D $y = -1$

II-1

20. Solve: $-6x + 9 = 21$

A -5
B -2
C 5
D 6

II-1

21. Solve for a: $-2(-3 - 5) = 3 - a$

A -13
B 19
C 13
D -19

II-1

22. Solve $3x^2 + x - 2 = 0$ by factoring.

   A $x = -2, 3$

   B $x = 1, 3$

   C $x = -\frac{2}{3}, 0$

   D $x = -1, \frac{2}{3}$

                        II-2

23. Solve: $x^2 + 2x - 35 = 0$

   A $x = -3, 5$

   B $x = -7, 5$

   C $x = -4, 6$

   D $x = -4, 9$

                        II-2

24. Solve $6x^2 - 24x - 270 = 0$ using the quadratic formula.

   A $x = -5, 9$

   B $x = 6, 15$

   C $x = -7, 5$

   D $x = -9, 3$

                        II-2

25. Where do these two equations intersect?

$$3x + y = 3$$
$$2y - x = -8$$

   A $(0, 1)$

   B $(0, 3)$

   C $(2, 3)$

   D $(2, -3)$

                        II-3

26. What is the solution to the following system of equations?

$$2x + 4y = 40$$
$$x = 3y$$

   A $(6, 4)$

   B $(12, 4)$

   C $(4, 6)$

   D $(4, 12)$

                        II-3

27. Where does this system of equations intersect?

$$y = -x$$
$$3x + 2y = -4$$

   A $(-4, -4)$

   B $(4, -4)$

   C $(-4, 4)$

   D $(4, 4)$

                        II-3

28. What is the solution to the systems of equations shown below?

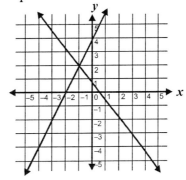

   A $(2, -1)$

   B $(-1, 2)$

   C $(-1, -1)$

   D $(2, 2)$

                        II-3

29. Solve the inequality $11x - 6 \geq -3x - 20$.

   A $x \geq -2$

   B $x \geq -1$

   C $x \leq 7$

   D $x \leq 3$

                        II-4

30. Solve the following inequality:
$$-3(4x + 5) > (5x + 6) + 13$$

   A $x < -\frac{14}{17}$

   B $x > -2$

   C $x > \frac{20}{11}$

   D $x < -2$

                        II-4

31. Find c: $\dfrac{c}{-2} > -6$

   A $c > -12$
   B $c < 12$
   C $c > 12$
   D $c < 3$

II-4

32. Solve: $6x + 2(24) \ge 150$

   A $x \ge 17$
   B $x \ge 20$
   C $x \ge 21$
   D $x \ge 25$

II-4

33. Is $y^2 = x$ a function? Explain your reasoning.

   A Yes, because it passes the horizontal line test.
   B Yes, because it passes the vertical line test.
   C No, because it fails horizontal line test.
   D No, because it fails the vertical line test.

III-1

34. Jittery Java sells bags of coffee by the pound. The table below shows the price of Jittery Java's coffee.

Coffee (x)	Price (y)
2 lbs	$6.50
3 lbs	$9.50
4 lbs	$12.50
5 lbs	$15.50
6 lbs	$18.50

Which equation best represents the data in the table?

   A $y = 0.5x + 3$
   B $y = 3.5x$
   C $y = 3x + 0.5$
   D $y = 3x - 0.5$

III-1

35. Which of these mapping is NOT a function?

A
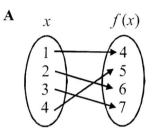

B
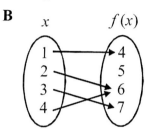

C
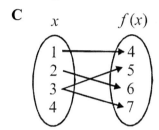

D
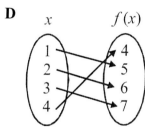

III-1

36. Sriram has written a computer program that randomly substitutes a value of x into a relation and produces one value for y. So far his computer program has generated the following data points:

$(5, 2), (10, -3), (17, -4), (37, 6), (65, -8)$

Which of the following data points produced by the computer program would show that the relation it is using is not a function?

   A $(1, 0)$
   B $(2, 1)$
   C $(26, -5)$
   D $(37, -6)$

III-1

37. $f(x) = -x + 1$; What is $f(-4)$?

   A -3
   B -5
   C 3
   D 5

III-2

38. What is the range of this function?
$\{(-1, 4), (2, -2), (3, 8), (5, -1)\}$

   A $\{-2, -1, 4, 8\}$
   B $\{-1, 2, 3, 5\}$
   C $\{-2, 8\}$
   D $\{-1, 5\}$

III-2

39. What is the range of the function shown on the graph?

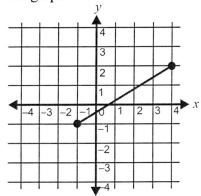

   A $0 \leq y \leq 4$
   B $0 \leq y \leq 5$
   C $-1 \leq y \leq 2$
   D $-1 \leq y \leq 4$

III-2

40. Find the range of the following function for the domain $\{-2, -1, 0, 3\}$.

$y = 3x - 3$

   A $\{-6, -3, 0, 6\}$
   B $\{-9, -6, -3, 6\}$
   C $\{-6, 0, 3, 9\}$
   D $\{-9, 0, 3, 9\}$

III-2

41. Using the formula $A = \pi r^2$, Sally calculated the area of a round wading pool with a diameter of 10 feet to be 314 square feet. Maybelle believes this is not correct because

   A $3 \times 5 = 15$.
   B $3 \times 10 = 30$.
   C $3 \times 5 \times 5 = 75$.
   D $3 \times 20 \times 20 = 1200$.

IV-1

42. Emily needs to make a glass case with the following measurements:

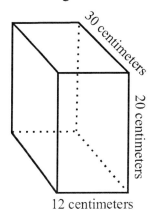

12 centimeters

How many square centimeters of glass would it take to construct the case enclosed on all sides?

   A 60 square centimeters
   B 612 square centimeters
   C 2,400 square centimeters
   D 6,200 square centimeters

IV-1

43. Alexandria wants to locate the midpoint of a line segment with endpoints $(-3, -2)$ and $(6, -4)$. What are the coordinates of the midpoint?

   A $(1.5, -3)$
   B $(4.5, -3)$
   C $(4.5, -6)$
   D $(3, -6)$

IV-2

44. What are the coordinates of the midpoint of $\overline{AB}$ if $A = (0, -6)$ and $B = (-2, 1)$?

A $(-2, -5)$
B $(-1, -3.5)$
C $(1, -3.5)$
D $(-1, -2.5)$

IV-2

45. What is the volume of the following oil tank? Round your answer to the nearest hundredth.

Use the formula $V = \pi r^2 h$, $\pi = 3.14$.

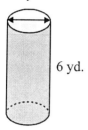
2 yd.

6 yd.

A 18.84 yd^3
B 37.68 yd^3
C 44.48 yd^3
D 75.36 yd^3

IV-1

46. Using the distance formula, what is the length of $\overline{AB}$?

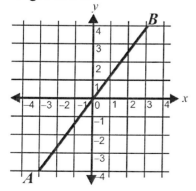

A $\sqrt{14}$
B 8
C 10
D 20

IV-2

47. Geoff made a circular flower garden in his yard that is 3 feet in diameter. He wants to buy edging that will keep the grass from growing in with the flowers. How many feet of edging will he need to buy to go around the outside of the circular flower garden? Use $\pi = 3.14$.

A 4.71 feet
B 6.00 feet
C 9.42 feet
D 18.84 feet

IV-1

48. What is the slope of the line in the graph below?

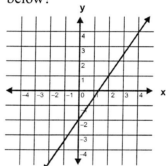

A 3
B -3
C $\frac{2}{3}$
D $\frac{3}{2}$

IV-2

49. Which of these equations represents the graph below?

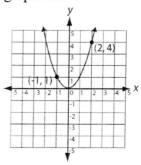

A $y = \sqrt{x}$
B $y = |x|$
C $y = x$
D $y = x^2$

V-1,4

50. Which is the graph of $f(x) = \sqrt{x}$?

A

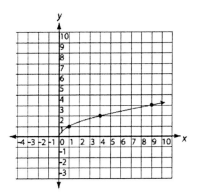

B

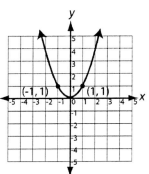

C

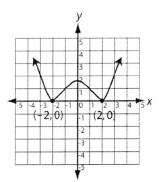

D

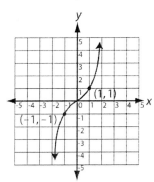

V-1,4

51. Which of the following is the graph of the equation $y = x - 3$?

A

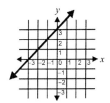

B

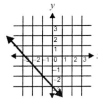

C

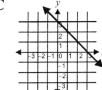

D

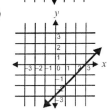

V-1,4

52. Which of the following equations is represented by the graph?

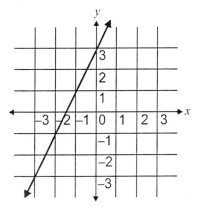

A $y = -3x + 3$
B $y = -\frac{1}{3}x + 3$
C $y = 3x - 3$
D $y = 2x + 3$

V-1,4

7

53. Which graph represents the equation $2y = 3x + 4$?

A

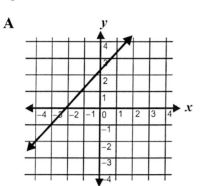

B

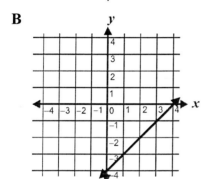

C

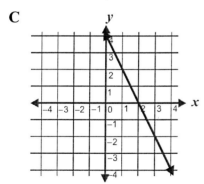

D

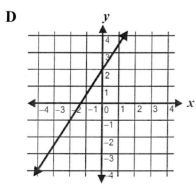

54. Which graph represents the equation $-12x + 6y = -6$?

A

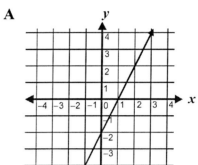

B

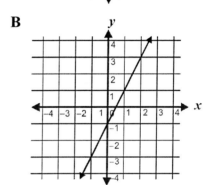

C

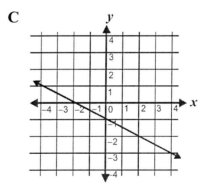

D

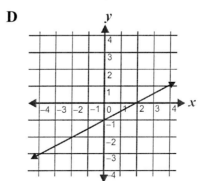

V-1,4

V-1,4

8

55. Which of the following graphs shows a line with a slope of $-\frac{1}{3}$ that passes through the point $(0, -1)$?

A

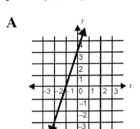

B

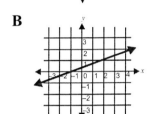

C

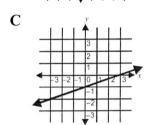

D

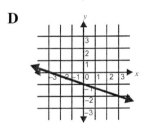

56. Which of these graphs represents the solution of $x > 2$?

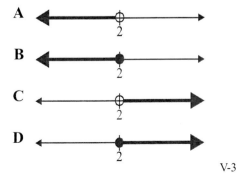

57. What graph shows a line that has a slope of -1 and a y-intercept of $(0, 1)$?

A

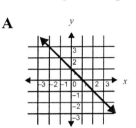

B

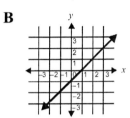

C

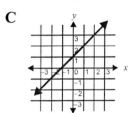

D

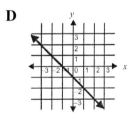

58. Which of these graphs represents the solution of $-8 \geq -4x > -36$?

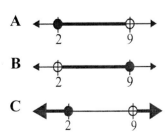

59. Which graph represents a line containing the points $(-2, -3)$ and $(1, 4)$?

A

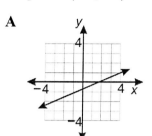

B

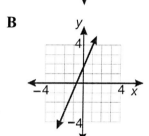

C

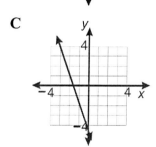

D

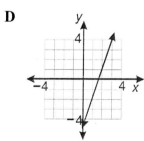

V-2

60. Andrea has 10 more jellybeans than her friend Chelsea, but she has half as many as Rebecca. Which expression below best describes Rebecca's jelly beans?

A $R = 2C + 20$
B $R = C + 10$
C $R = A + \frac{1}{2}C$
D $R = 2A + 10$

VI-1

61. Which figure below illustrates the graph of the equation with a x-intercept of 1 and y-intercept of -1?

A

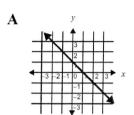

B

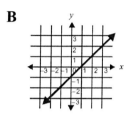

C

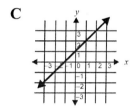

D

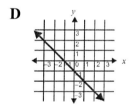

V-2

62. Which of these graphs represents the solution of $2x > -3$?

A

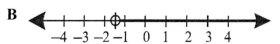

B

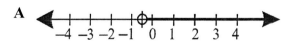

C

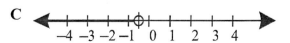

D

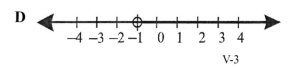

V-3

10

63. Which of these graphs represents the solution of $-3 \leq x < 12$?

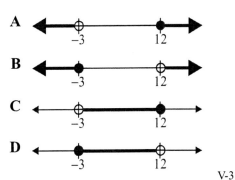

A

B

C

D

V-3

64. Look at the graph below.

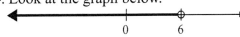

Which inequality is represented on the number line?

A $x > 6$
B $x \geq 6$
C $x < 6$
D $x \leq 6$

VI-1

65. What is the equation of the line that includes the point $(4, -3)$ and has a slope of -2?

A $y = -2x - 5$
B $y = -2x - 2$
C $y = -2x + 5$
D $y = 2x - 5$

VI-1

66. Translate the following sentence into an algebraic equation:
Two times the sum of a number, x, and 6 is 14.

A $2x + 6 = 14$
B $2x - 6 = 14$
C $2(x + 6) = 14$
D $2(x - 6) = 14$

VI-1

67. Which of the following equations is represented by the graph?

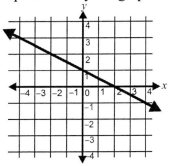

A $y = -\frac{1}{2}x + 1$
B $y = -2x + 1$
C $y = \frac{1}{2}x + 1$
D $y = 2x + 1$

VI-1

68. The biology club plans to rent a river boat for an outing. The cost will be $250.00, plus $25.00 for each hour (h) the boat is used. Which statement represents the total cost (c) of renting the boat?

A $c = 275h$
B $c = 250 + 25h$
C $c = 275h + 25$
D $c = 275 + 25h$

VI-1

69. Quadrilateral $ABCD$ is a parallelogram. What is the measure of $\angle ECF$?

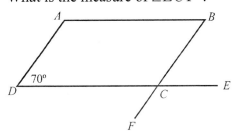

A $70°$
B $110°$
C $130°$
D $140°$

VII-1

70. Which two angles are supplementary?

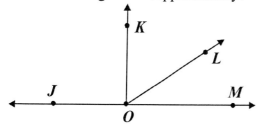

   A ∠JOL and ∠KOM
   B ∠LOM and ∠KOL
   C ∠JOK and ∠KOL
   D ∠KOM and ∠JOK

VII-1

71. In the figure below, the measure of ∠4 is 130°. What is the sum of the measures of ∠2 and ∠3?

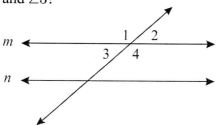

   A 50°
   B 60°
   C 80°
   D 100°

VII-1

72. Triangle ABC, shown in the diagram below, is an isosceles triangle.

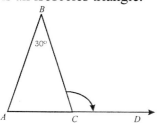

What is the measure, in degrees, of ∠BCD?

   A 75°
   B 105°
   C 150°
   D 165°

VII-1

73. The stage crew will be making flat panels to form a wall on the stage. The panels must fit through the stage door as shown below.

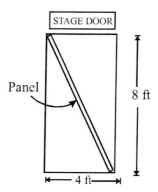

What is the maximum width of a panel?

   A $2\sqrt{5}$
   B $4\sqrt{5}$
   C $16\sqrt{5}$
   D $\sqrt{12}$

VII-2

74. A 25-foot ladder is leaning against a building. The base of the ladder is 7 feet from the base of the building. How high up the building does the ladder reach?

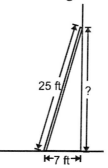

   A 21 feet
   B 22 feet
   C 23 feet
   D 24 feet

VII-2

75. An excursion boat traveled from the Ferry Dock to Shelter Cove. How far did it travel?

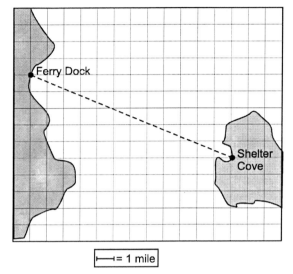

= 1 mile

A 12 miles
B 13 miles
C 14 miles
D 15 miles

VII-2

76. Find the length of the hypotenuse, h, of the following triangle.

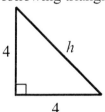

A 16
B $4\sqrt{2}$
C 6
D 32

VII-2

77. What is the radius of the base of a cylinder whose volume is $5,086.8$ cubic yards and height is 20 yards?

A 9 yards
B 18 yards
C 14 yards
D 6 yards

VII-4

78. Rosa has purchased a larger delivery van for her business.

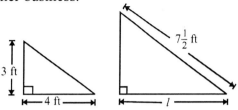

She intends to use a similar, larger logo sign on the new van. What will be the length (l) of the new logo sign?

A 5 feet
B $5\frac{1}{4}$ feet
C 6 feet
D 10 feet

VII-3

79. Below is a drawing of a farm plot:

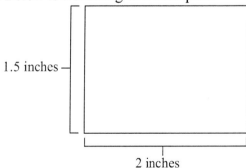

Scale: 1 inch = 0.75 miles
What is the perimeter of this farm plot?

A $2\frac{1}{4}$ miles
B 3 miles
C $5\frac{1}{4}$ miles
D 7 miles

VII-3

80. What is the area of a circle with $r = 3x - 1$?

A $(6x - 2)\pi$
B $(12x - 4)\pi$
C $(6x^2 + 3x + 1)\pi$
D $(9x^2 - 6x + 1)\pi$

VII-4

81.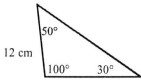

Which triangle is similar to the triangle above?

A
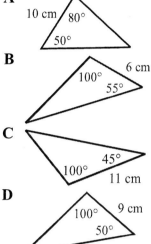
10 cm 80°
50°

B
6 cm
100°
55°

C
45°
100°
11 cm

D
100° 9 cm
50°

VII-3

82. How many square feet is the area of the great room below?

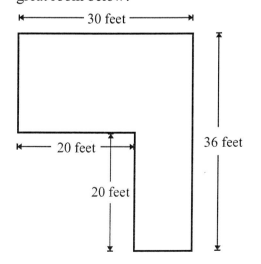

← 30 feet →

← 20 feet →

20 feet

36 feet

A 480 feet²
B 680 feet²
C 720 feet²
D 1080 feet²

VII-4

83. The two trapezoids below are similar.

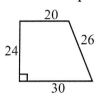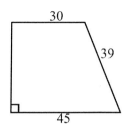
20
24   26
30

30
39
45

What is the height of the larger trapezoid?

A 16
B 30
C 32
D 36

VII-3

84. Find the perimeter of the rectangle below.

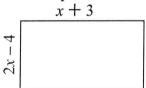
$x + 3$
$2x - 4$

A $2x^2 - 2x - 1$
B $3x - 1$
C $2x^2 + 2x - 12$
D $6x - 2$

VII-4

85. Jeff's Wholesale Auto Dealership holds an auction once a month. What is the average number of cars sold at the auction for the past 6 months?

Month 1: 288 Cars   Month 4: 558 Cars
Month 2: 432 Cars   Month 5: 366 Cars
Month 3: 330 Cars   Month 6: 258 Cars

A 318.9 cars
B 350 cars
C 446.6 cars
D 372 cars

VII-5

86. A neighborhood surveyed the times of day people water their lawns and tallied the data below.

Time	Tally
midnight - 3:59 a.m.	II
4:00 a.m. - 7:59 a.m.	JHT I
8:00 a.m. - 11:59 a.m.	JHT IIII
noon - 3:59 p.m.	JHT
4:00 p.m. - 7:59 p.m.	JHT JHT
8:00 p.m. - 11:59 p.m.	JHT III

If you wanted to find which was the most popular time of day to water the lawn, it would be best to find the _____ of data.

A mean
B median
C range
D mode

VII-5

87. What is the median of the following set of data?
33, 31, 35, 24, 38, 30

A 32
B 31
C 30
D 29

VII-5

88. Jack has a bag of marbles containing 4 blue striped, 7 red striped, 3 white, and 4 brown. If Jack reaches in the bag keeping his eyes closed, what is the probability that Jack will get a white or brown marble?

A $\frac{5}{18}$
B $\frac{7}{18}$
C $\frac{4}{18}$
D $\frac{2}{3}$

VII-6

89. What is the range of the data below?
3.2, 5.4, 1.4, 9.6, 4.1, 3.7, 5.4

A 8.2
B 5.4
C 7.6
D 4.7

VII-5

90. Casey, our pet iguana, whacked his tail into a pile of marbles. One of the marbles went under the couch, never to be seen again. There were 6 red marbles, 11 orange marbles, 4 blue marbles, and 7 multicolored marbles. What is the probability the one missing is orange?

A 11 out of 28
B 1 out of 28
C 17 out of 28
D 27 out of 28

VII-6

91. What is the probability of landing on a green segment using the spinner below one time? Express your answer in fraction form.

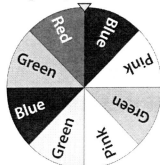

A $\frac{1}{4}$
B $\frac{3}{8}$
C $\frac{1}{8}$
D $\frac{3}{4}$

VII-6

92. What is the probability that the spinner will land on a shaded section or the number 4?

A $\frac{1}{6}$

B $\frac{1}{3}$

C $\frac{1}{2}$

D $\frac{2}{3}$

VII-6

93. A recipe for 32 ounces of lemonade calls for 4 ounces of lemon juice. Janet wants to make 120 ounces of lemonade. Which proportion below should she use to find the amount of lemon juice needed?

A $\frac{32}{120} = \frac{x}{4}$

B $\frac{x}{32} = \frac{4}{120}$

C $\frac{32}{4} = \frac{x}{120}$

D $\frac{4}{32} = \frac{x}{120}$

VII-7

94. Each lap around the lake is $\frac{4}{5}$ of a mile. How many miles did Teri run if she ran $4\frac{1}{2}$ laps?

A $3\frac{3}{5}$

B $3\frac{1}{5}$

C $4\frac{5}{8}$

D $5\frac{5}{8}$

VII-7

95. A pine tree casts a shadow 9 feet long. At the same time, a rod measuring 4 feet casts a shadow 1.5 feet long. How tall is the pine tree?

A 24

B 3.375

C 6

D 36

VII-7

96. If $y = 6$ and $x = 4$, what is the value of y when $x = 8$ if this represents a direct variation?

A 4

B 12

C 3

D 8

VII-7

97. Linda is starting a pet-grooming business to pay back the $250.75 she borrowed from her mom. She figured out that the supplies are going to cost $81.95. She plans to groom 8 dogs per week. How much will Linda have to charge for each dog in order to cover her expenses and pay her mom back in 3 weeks?

A $34.10

B $10.45

C $15.63

D $13.86

VII-8

98. Together Jack, Alicia, and Ashley spent $29.00 on a movie. Alicia gave three more dollars than Jack. Ashley gave twice as much as Alicia. How much money did Ashley give?

A $10.00

B $11.00

C $13.00

D $16.00

VII-8

99. Fido gets 2 doggy treats every time he sits and 4 doggy treats when he rolls over on command. Throughout the week, he has sat 6 times as often as he has rolled over. In total, he has earned 80 doggy treats. How many times has Fido sat?

   A 4
   B 5
   C 24
   D 30

   VII-8

100. Brad and Kyle went to O'Donald's to get a snack after football practice. The table below shows what they bought and the amount they paid.

	Fries	Drinks	Total Cost
Brad	2	2	$4.02
Kyle	3	1	$4.25

What is the cost of 1 order of French fries?

   A $0.89
   B $1.12
   C $1.24
   D $1.49

   VII-8

Evaluation Chart for the Diagnostic Mathematics Test

Directions: On the following chart, circle the question numbers that you answered incorrectly. Then turn to the appropriate topics (listed by chapters), read the explanations, and complete the exercises. Review the other chapters as needed. Finally, complete *Passing the Alabama High School Graduation Exam in Mathematics* Practice Tests for further review.

		Questions	Pages
Chapter 1:	Introduction to Algebra	1, 2, 3, 60, 66, 68	19–30
Chapter 2:	Solving One-Step Equations and Inequalities	31, 56, 62, 63, 64	31–41
Chapter 3:	Solving Multi-Step Equations and Inequalities	4, 18, 19, 20, 21, 29, 30, 32, 58	42–54
Chapter 4:	Ratios and Proportions	93, 94, 95, 96	55–61
Chapter 5:	Algebra Word Problems	97, 98, 99	62–77
Chapter 6:	Polynomials	5, 6, 7, 8, 9, 10, 11, 12	78–90
Chapter 7:	Factoring	13, 14, 16, 17	91–104
Chapter 8:	Solving Quadratic Equations	15, 22, 23, 24	105–112
Chapter 9:	Graphing and Writing Equations	43, 44, 46, 48, 51, 52, 53, 54, 55, 57, 59, 61, 65, 67	113–129
Chapter 10:	Systems of Equations	25, 26, 27, 28, 100	130–141
Chapter 11:	Relations and Functions	33, 34, 35, 36, 37, 38, 39, 40, 49, 50	142–162
Chapter 12:	Statistics	85, 86, 87, 89	163–172
Chapter 13:	Probability	88, 90, 91, 92	173–180
Chapter 14:	Angles	69, 70, 71	181–190
Chapter 15:	Triangles	72, 73, 74, 75, 76, 78, 81	191–202
Chapter 16:	Plane Geometry	41, 47, 79, 80, 82, 83, 84	203–220
Chapter 17:	Solid Geometry	42, 45, 77	221–230

Chapter 1
Introduction to Algebra

This chapter covers the following Alabama objectives and standards in mathematics:

	Objective(s)
Standard I	1
Standard VI	1

1.1 Algebra Vocabulary

Vocabulary Word	Example	Definition
variable	$4x$ (x is the variable)	a letter that can be replaced by a number
coefficient	$4x$ (4 is the coefficient)	a number multiplied by a variable or variables
term	$5x^2 + x - 2$ ($5x^2$, x, and -2 are terms)	numbers or variables separated by $+$ or $-$ signs
constant	$5x + 2y + 4$ (4 is a constant)	a term that does not have a variable
degree	$4x^2 + 3x - 2$ (the degree is 2)	the largest power of a variable in an expression
leading coefficient	$4x^2 + 3x - 2$ (4 is the leading coefficient)	the number multiplied by the term with the highest power
sentence	$2x = 7$ or $5 \leq x$	two algebraic expressions connected by $=, \neq, <, >, \leq, \geq$, or $\approx$
equation	$4x = 8$	a sentence with an equal sign
inequality	$7x < 30$ or $x \neq 6$	a sentence with one of the following signs: $\neq, <, >, \leq, \geq$, or $\approx$
base	6^3 (6 is the base)	the number used as a factor
exponent	6^3 (3 is the exponent)	the number of times the base is multiplied by itself

1.2   Substituting Numbers for Variables

These problems may look difficult at first glance, but they are very easy. Simply replace the variable with the number the variable is equal to, and solve the problems.

Example 1:   In the following problems, substitute 10 for a.

Problem	**Calculation**	**Solution**
1. $a + 1$	Simply replace the a with 10. $10 + 1$	11
2. $17 - a$	$17 - 10$	7
3. $9a$	This means multiply. 9×10	90
4. $\dfrac{30}{a}$	This means divide. $30 \div 10$	3
5. a^3	$10 \times 10 \times 10$	1000
6. $5a + 6$	$(5 \times 10) + 6$	56

Note: Be sure to do all multiplying and dividing before adding and subtracting.

Example 2:   In the following problems, let $x = 2$, $y = 4$, and $z = 5$.

Problem	**Calculation**	**Solution**
1. $5xy + z$	$5 \times 2 \times 4 + 5$	45
2. $xz^2 + 5$	$2 \times 5^2 + 5 = 2 \times 25 + 5$	55
3. $\dfrac{yz}{x}$	$(4 \times 5) \div 2 = 20 \div 2$	10

In the following problems, $t = 7$. Solve the problems.

1. $t + 3 =$

2. $18 - t =$

3. $\dfrac{21}{t} =$

4. $3t - 5 =$

5. $t^2 + 1 =$

6. $2t - 4 =$

7. $9t \div 3 =$

8. $\dfrac{t^2}{7} =$

9. $5t + 6 =$

10. $\dfrac{(t^2 - 7)}{6} =$

11. $4t + 5t =$

12. $\dfrac{6t}{3} =$

In the following problems $a = 4$, $b = -2$, $c = 5$, and $d = 10$. Solve the problems.

13. $4a + 2c =$

14. $3bc - d =$

15. $\dfrac{ac}{d} =$

16. $d - 2a =$

17. $a^2 - b =$

18. $abd =$

19. $5c - ad =$

20. $cd + bc =$

21. $\dfrac{6b}{a} =$

22. $9a + b =$

23. $5 + 3bc =$

24. $d^2 + d + 1 =$

1.3   Understanding Algebra Word Problems

The biggest challenge to solving word problems is figuring out whether to add, subtract, multiply, or divide. Below is a list of key words and their meanings. This list does not include every situation you might see, but it includes the most common examples.

Words Indicating Addition	Example	Add
and	6 **and** 8	$6 + 8$
increased	The original price of $15 **increased** by $5.	$15 + 5$
more	3 coins and 8 **more**	$3 + 8$
more than	Josh has 10 points. Will has 5 **more than** Josh.	$10 + 5$
plus	8 baseballs **plus** 4 baseballs	$8 + 4$
sum	the **sum** of 3 and 5	$3 + 5$
total	the **total** of 10, 14, and 15	$10 + 14 + 15$

Words Indicating Subtraction	Example	Subtract
decreased	$16 **decreased** by $5	$16 - 5$
difference	the **difference** between 18 and 6	$18 - 6$
less	14 days **less** 5	$14 - 5$
less than	Jose completed 2 laps **less than** Mike's 9.	*$9 - 2$
left	Ray sold 15 out of 35 tickets. How many did he have **left**?	*$35 - 15$
lower than	This month's rainfall is 2 inches **lower than** last month's rainfall of 8 inches.	*$8 - 2$
minus	15 **minus** 6	$15 - 6$

* In subtraction word problems, you cannot always subtract the numbers in the order that they appear in the problem. Sometimes the first number should be subtracted from the last. You must read each problem carefully.

Words Indicating Multiplication	Example	Multiply
double	Her $1000 profit **doubled** in in a month.	1000×2
half	**Half** of the $600 collected went to charity.	$\frac{1}{2} \times 600$
product	the **product** of 4 and 8	4×8
times	Li scored 3 **times** as many points as Ted who only scored 4.	3×4
triple	The bacteria **tripled** its original colony of 10,000 in just one day.	$3 \times 10,000$
twice	Ron has 6 CDs. Tom has **twice** as many.	2×6

Words Indicating Division	Example	Divide
divide into, by, or among	The group of 70 **divided into** 10 teams	$70 \div 10$ or $\frac{70}{10}$
quotient	the **quotient** of 30 and 6	$30 \div 6$ or $\frac{30}{6}$

Match the phrase with the correct algebraic expression below. The answers will be used more than once.

A. $y - 2$

B. $2y$

C. $y + 2$

D. $\dfrac{y}{2}$

E. $2 - y$

1. 2 more than y	5. the quotient of y and 2	9. y decreased by 2
2. y divided into 2	6. y increased by 2	10. y doubled
3. 2 less than y	7. 2 less y	11. 2 minus y
4. twice y	8. the product of 2 and y	12. the total of 2 and y

Now practice writing parts of algebraic expressions from the following word problems.

Example 3:   the product of 3 and a number, t Answer: $3t$

13. 3 less than x	23. bacteria culture, b, doubled
14. y divided among 10	24. triple John's age, y
15. the sum of t and 5	25. a number, n, plus 4
16. n minus 14	26. quantity, t, less 6
17. 5 times k	27. 18 divided by a number, x
18. the total of z and 12	28. n feet lower than 10
19. double the number b	29. 3 more than p
20. x increased by 1	30. the product of 4 and m
21. the quotient t and 4	31. a number, y, decreased by 20
22. half of a number, y	32. 5 times as much as x

1.4   Setting Up Algebra Word Problems

So far, you have seen only the first part of algebra word problems. To complete an algebra problem, an equal sign must be added. The words **"is"** or **"are"** as well as **"equal(s)"** signal that you should add an equal sign.

Example 4:
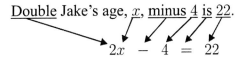

Double Jake's age, x, minus 4 is 22.

$$2x - 4 = 22$$

Translate the following word problems into algebra problems. DO NOT find the solutions to the problems yet.

1. Triple the original number, n, is $2,700$.

2. The product of a number, y, and 5 is equal to 15.

3. Four times the difference of a number, x, and 2 is 20.

4. The total, t, divided into 5 groups is 45.

5. The number of parts in inventory, p, minus 54 parts sold today is 320.

6. One-half an amount, x, added to $50 is $262

7. One hundred seeds divided by 5 rows equals n number of seeds per row.

8. A number, y, less than 50 is 82.

9. His base pay of $200 increased by his commission, x, is $500.

10. Seventeen more than half a number, h, is 35.

11. This month's sales of $2,300 are double January's sales, x.

12. The quotient of a number, w, and 4 is 32.

13. Six less a number, d, is 12.

14. Four times the sum of a number, y, and 10 is 48.

15. We started with x number of students. When 5 moved away, we had 42 left.

16. A number, b, divided into 36 is 12.

1.5   Changing Algebra Word Problems to Algebraic Equations

Example 5:   There are 3 people who have a total weight of 595 pounds. Sally weighs 20 pounds less than Jessie. Rafael weighs 15 pounds more than Jessie. How much does Jessie weigh?

Step 1:   Notice everyone's weight is given in terms of Jessie. Sally weighs 20 pounds less than Jessie Rafael weighs 15 pounds more than Jessie. First, we write everyone's weight in terms of Jessie, j.

$$\begin{aligned} \text{Jessie} &= j \\ \text{Sally} &= j - 20 \\ \text{Rafael} &= j + 15 \end{aligned}$$

Step 2:   We know that all three together weigh 595 pounds. We write the sum of everyone's weight equal to 595.

$$j + j - 20 + j + 15 = 595$$

We will learn to solve these problems in chapter 3.

Change the following word problems to algebraic equations.

1. Fluffy, Spot, and Shampy have a combined age in dog years of 91. Spot is 14 years younger than Fluffy. Shampy is 6 years older than Fluffy. What is Fluffy's age, f, in dog years?

2. Jerry Marcosi puts 5% of the amount he makes per week into a retirement account, r. He is paid $11.00 per hour and works 40 hours per week for a certain number of weeks, w. Write an equation to help him find out how much he puts into his retirement account.

3. A furniture store advertises a 40% off liquidation sale on all items. What would the sale price (p) be on a $2530 dining room set?

4. Kyle Thornton buys an item which normally sells for a certain price, x. Today the item is selling for 25% off the regular price. A sales tax of 6% is added to the equation to find the final price, f.

5. Tamika Francois runs a floral shop. On Tuesday, Tamika sold a total of $600 worth of flowers. The flowers cost her $100, and she paid an employee to work 8 hours for a given wage, w. Write an equation to help Tamika find her profit, p, on Tuesday.

6. Sharice is a waitress at a local restaurant. She makes an hourly wage of $3.50, plus she receives tips. On Monday, she works 6 hours and receives tip money, t. Write an equation showing what Sharice makes on Monday, y.

7. Jenelle buys x shares of stock in a company at $34.50 per share. She later sells the shares at $40.50 per share. Write an equation to show how much money, m, Jenelle has made.

1.6 Substituting Numbers in Formulas

Example 6: Area of a parallelogram: $A = b \times h$
Find the area of the parallelogram if $b = 20$ cm and $h = 10$ cm.

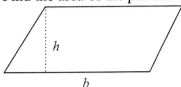

Step 1: Copy the formula with the numbers given in place of the letter in the formula.
$A = 20 \times 10$

Step 2: Solve the problem. $A = 20 \times 10 = 200$. Therefore, $A = 200$ cm^2.

Solve the following problems using the formulas given.

1. The volume of a rectangular pyramid is determined by using the following formula:
$$V = \frac{lwh}{3}$$
Find the volume of the pyramid if $l = 6$ in, $w = 6$ in, and $h = 11$ in.

2. Find the volume of a cone with a radius of 30 inches and a height of 60 inches using the formula:
$V = \frac{1}{3}\pi r^2 h$ $\pi = 3.14$

3. Lumber is measured by the following formula:
Number of board feet $= \dfrac{LWT}{12}$
Find the number of board feet if $L = 14$ feet, $W = 8$ feet, and $T = 6$ feet.

4. The perimeter of a square is figured by the formula $P = 4s$.
Find the perimeter if $s = 6$.

5. What is the circumference of a circle with a diameter of 8 cm?
$C = \pi d$ $\pi = 3.14$

6. Find the area of the trapezoid

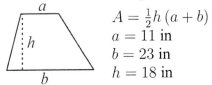

$A = \frac{1}{2}h(a+b)$
$a = 11$ in
$b = 23$ in
$h = 18$ in

7. Find the volume of a sphere with a radius of 6 cm. $\pi = 3.14$

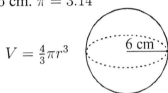

$V = \frac{4}{3}\pi r^3$

8. Find the area of the following ellipse given by the equation: $A = \pi ab$

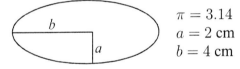

$\pi = 3.14$
$a = 2$ cm
$b = 4$ cm

9. The formula for changing from degrees Fahrenheit to degrees Celsius is:
$$C = \frac{5(F - 32)}{9}$$
If it is 68°F outside, how many degrees Celsius is it?

10. Find the volume. $V = \frac{4}{3}\pi r^3$ $\pi = 3.14$

11. Louise has a cone-shaped mold to make candles. The diameter of the base is 10 cm, and it is 13 cm tall. How many cubic centimeters of liquid wax will it hold?
$\pi = 3.14$
$V = \frac{1}{3}\pi r^2 h$

1.7   Order of Operations

In long math problems with $+$, $-$, $\times$, $\div$, $()$, and exponents in them, you have to know what to do first. Without following the same rules, you could get different answers. If you will memorize the silly sentence, Please Excuse My Dear Aunt Sally, you can memorize the order you must follow.

<u>P</u>lease "**P**" stands for parentheses. You must get rid of parentheses first.
Examples: $3(1+4) = 3 \times 5 = 15$
$\qquad\qquad\quad 6(10-6) = 6 \times 4 = 24$

<u>E</u>xcuse "**E**" stands for exponents. You must eliminate exponents next.
Example: $4^2 = 4 \times 4 = 16$

<u>M</u>y <u>D</u>ear "**M**" stands for multiply. "**D**" stands for divide. Start on the left of the equation and perform all multiplications and divisions in the order in which they appear.

<u>A</u>unt <u>S</u>ally "**A**" stands for add. "**S**" stands for subtract. Start on the left and perform all additions and subtractions in the order they appear.

Example 7: $12 \div 2(6-3) + 3^2 - 1$

Please Eliminate **parentheses**. $6 - 3 = 3$ so now we have $12 \div 2 \times 3 + 3^2 - 1$

Excuse Eliminate **exponents**. $3^2 = 9$ so now we have $12 \div 2 \times 3 + 9 - 1$

My Dear **Multiply** and **divide** next in order from left to right. $12 \div 2 = 6$ then $6 \times 3 = 18$

Aunt Sally Last, we **add** and **subtract** in order from left to right. $18 + 9 - 1 = 26$

Simplify the following problems.

1. $6 + 9 \times 2 - 4$

2. $3(4+2) - 6^2$

3. $3(6-3) - 2^3$

4. $49 \div 7 - 3 \times 3$

5. $10 \times 4 - (7-2)$

6. $2 \times 3 \div 6 \times 4$

7. $4^3 \div 8(4+2)$

8. $7 + 8(14-6) \div 4$

9. $(2+8-12) \times 4$

10. $4(8-13) \times 4$

11. $8 + 4^2 \times 2 - 6$

12. $3^2(4+6) + 3$

13. $(12-6) + 27 \div 3^2$

14. $82^0 - 1 + 4 \div 2^2$

15. $1 - (2-3) + 8$

16. $12 - 4(7-2)$

17. $18 \div (6+3) - 12$

18. $10^2 + 3^3 - 2 \times 3$

19. $4^2 + (7+2) \div 3$

20. $7 \times 4 - 9 \div 3$

When a problem has a fraction bar, simplify the top of the fraction (numerator) and the bottom of the fraction (denominator) separately using the rules for order of operations. You treat the top and bottom as if they were separate problems. Then reduce the fraction to lowest terms.

Example 8: $\dfrac{2(4-3)-6}{5^2+3(2+1)}$

Please	Eliminate **parentheses**. $(4-3)=1$ and $(2+1)=3$	$\dfrac{2\times1-6}{5^2+3\times3}$
Excuse	Eliminate **exponents**. $5^2=25$	$\dfrac{2\times1-6}{25+3\times3}$
My Dear	**Multiply** and **divide** in the numerator and denominator separately. $3\times3=9$ and $2\times1=2$	$\dfrac{2-6}{25+9}$
Aunt Sally	**Add** and **subtract** in the numerator and denominator separately. $2-6=-4$ and $25+9=34$	$\dfrac{-4}{34}$

Now reduce the fraction to lowest terms. $\dfrac{-4}{34}=\dfrac{-2}{17}$

Simplify the following problems.

1. $\dfrac{2^2+4}{5+3(8+1)}$

2. $\dfrac{8^2-(4+11)}{4^2-3^2}$

3. $\dfrac{5-2(4-3)}{2(1-8)}$

4. $\dfrac{10+(2-4)}{4(2+6)-2^2}$

5. $\dfrac{3^3-8(1+2)}{-10-(3+8)}$

6. $\dfrac{(9-3)+3^2}{-5-2(4+1)}$

7. $\dfrac{16-3(10-6)}{(13+15)-5^2}$

8. $\dfrac{(2-5)-11}{12-2(3+1)}$

9. $\dfrac{7+(8-16)}{6^2-5^2}$

10. $\dfrac{16-(12-3)}{8(2+3)-5}$

11. $\dfrac{-3(9-7)}{7+9-2^3}$

12. $\dfrac{4-(2+7)}{13+(6-9)}$

13. $\dfrac{5(3-8)-2^2}{7-3(6+1)}$

14. $\dfrac{3(3-8)+5}{8^2-(5+9)}$

15. $\dfrac{6^2-4(7+3)}{8+(9-3)}$

Chapter 1 Review

Solve the following problems using $x = 2$.

1. $3x + 4 =$

2. $\dfrac{6x}{4} =$

3. $x^2 - 5 =$

4. $\dfrac{x^3 + 8}{2} =$

5. $12 - 3x =$

6. $x - 5 =$

7. $-5x + 4 =$

8. $9 - x =$

9. $2x + 2 =$

Solve the following problems. Let $w = -1, y = 3, z = 5$.

10. $5w - y =$

11. $wyz + 2 =$

12. $z - 2w =$

13. $\dfrac{3z + 5}{wz} =$

14. $-2y + 3 =$

15. $25 - 2yz =$

16. $\dfrac{6w}{y} + \dfrac{z}{w} =$

17. $4w - (yw) =$

18. $7y - 5z =$

Simplify the following problems using the correct order of operations.

19. $10 \div (-1 - 4) + 2$

20. $5 + (2)(4 - 1) \div 3$

21. $5 - 5^2 + (2 - 4)$

22. $(8 - 10) \times (5 + 3) - 10$

23. $\dfrac{10 + 5^2 - 3}{2^2 + 2(5 - 3)}$

24. $1 - (9 - 1) \div 2$

25. $\dfrac{5(3 - 6) + 3^2}{4(2 + 1) - 6}$

26. $-4(6 + 4) \div (-2) + 1$

27. $12 \div (7 - 4) - 2$

28. $1 + 4^2 \div (3 + 1)$

29. $8 + (5)(3 - 5)$

For questions 30–32, write an equation to match the problem.

30. Calista earns $450 per week for a 40-hour work week plus $16.83 per hour for each hour of overtime after 40 hours. Write an equation that would be used to determine her weekly wages where w is her wages and v is the number of overtime hours worked.

31. Daniel purchased a 1-year CD, c, from a bank. He bought it at an annual interest rate of 6%. After 1 year, Daniel cashes in the CD. What is the total amount it is worth?

32. Omar is a salesman. He earns an hourly wage of $8.00 per hour, plus he receives a commission of 7% on the sales he makes. Write an equation which would be used to determine his weekly salary, w, where x is the number of hours worked, and y is the amount of sales for the week.

Answer each of the following questions.

33. Lumber is measured with the following formula: Number of board feet $= \dfrac{LWT}{12}$

    Find the number of board feet if $L = 12$ feet, $W = 6$ feet, and $T = \frac{1}{4}$ feet.

34. To convert from degrees Celsius to degrees Fahrenheit, use the following formula:

$$F = \frac{9C}{5} + 32$$

    If it is $15°C$ outside, what is the temperature in degrees Fahrenheit?

Chapter 1 Test

1. Solving the following expression using $a = 4$.

   $3a - 2$

   A 14
   B 10
   C 16
   D 12

2. Solving the following expression using $y = 3$.

   $8y \div 3$

   A 8
   B 21
   C 24
   D 72

3. Write the expression from the following word problem.

   A number divided by the sum of nine and two.

   A $\dfrac{x}{9 + 2}$

   B $\dfrac{9 + 2}{x}$

   C $\dfrac{x}{9 - 2}$

   D $\dfrac{x + 2}{9}$

4. A box has length of 20 inches, the height of 12 inches, and the width of 38 inches. What is the volume of the box using the formula $V = lwh$?

   A 70 inches
   B 240 inches
   C 9, 120 inches
   D 2, 912 inches

5. Solving the following expression using $x = 2$ and $y = 5$.

   $3x + 4y - 1$

   A 22
   B 13
   C 25
   D 10

6. Write the expression from the following word problem.

   Five less than x plus seven.

   A $5 - (x + 7)$
   B $(x + 5) - 7$
   C $(x + 7) - 5$
   D $(x - 7) + 5$

7. Write the expression from the following word problem.

   Fifteen minus a number, then divided by two equals eleven.

   A $\dfrac{15 - y}{2} = 11$

   B $11 - \dfrac{y}{2} = 15$

   C $15 - \dfrac{y}{2} = 11$

   D $2 - \dfrac{y}{15} = 11$

8. Write the equation from the following word problem.
   Sixteen times two plus a number equals seven.

   A $16 \times 2 + x = 7$
   B $16 \times 2 = 7x$
   C $16 \times (2 + x) = 7$
   D $16 + 2 \times x = 7$

9. Tom earns $500 per week before taxes are taken out. His employer takes out a total of 33% for state, federal, and Social Security taxes. Which expression below will help Tom figure his net pay?

   A $500 - 0.33$

   B $500 \div 0.33$

   C $500 + 0.33\,(500)$

   D $500 - 0.33\,(500)$

10. Rosa has to pay the first $100 of her medical expenses each year before she qualifies for her insurance company to begin paying. After paying the $100 "deductible," her insurance company will pay 80% of her medical expenses. This year, her total medical expenses came to $960.00. Which expression below shows how much her insurance company will pay?

    A $0.80\,(960 - 100)$

    B $100 + (960 \div 0.80)$

    C $960\,(100 - 0.80)$

    D $0.80\,(960 + 100)$

11. A plumber charges $45 per hour plus a $25.00 service charge. If a represents his total charges in dollars and b represents the number of hours worked, which formula below could the plumber use to calculate his total charges?

    A $a = 45 + 25b$

    B $a = 45 + 25 + b$

    C $a = 45b + 25$

    D $a = (45)(25) + b$

12. In 2009, Bell Computers informed its sales force to expect a 2.6% price increase on all computer equipment in the year 2010. A certain sales representative wanted to see how much the increase would be on a computer, c, that sold for $2200 in 2009. Which expression below will help him find the cost of the computer in the year 2010?

    A $0.26\,(2200)$

    B $2200 - 0.026\,(2200)$

    C $2200 + 0.026\,(2200)$

    D $0.026\,(2200) - 2200$

13. Simplify: $\dfrac{4 - (-3) \cdot (-2)}{(-8) - (-6)} =$

    A $\dfrac{1}{7}$

    B 1

    C 7

    D $\dfrac{-1}{7}$

14. Simplify: $\dfrac{3[(-6) \cdot (5) + 15]}{10 + (-5)} =$

    A -9

    B -45

    C 15

    D -72

Chapter 2
Solving One-Step Equations and Inequalities

HSGE

Mathematics

This chapter covers the following Alabama objectives and standards in mathematics:

	Objective(s)
Standard V	3
Standard VI	1

2.1 One-Step Algebra Problems with Addition and Subtraction

You have been solving algebra problems since second grade by filling in blanks. For example, $5 + \_\_ = 8$. The answer is 3. You can solve the same kind of problems using algebra. The problems only look a little different because the blank has been replaced with a letter. The letter is called a **variable**.

Example 1:   **Arithmetic** $5 + \_\_ = 14$
                   Algebra $5 + x = 14$

The goal in any algebra problem is to move all the numbers to one side of the equal sign and have the letter (called a **variable**) on the other side. In this problem the 5 and the "x" are on the same side. The 5 is added to x. To move it, do the **opposite** of **add**. The **opposite** of **add** is **subtract**, so subtract 5 from both sides of the equation. Now the problem looks like this:

$$\begin{array}{r} 5 + x = 14 \\ -5 \quad\quad -5 \\ \hline x = 9 \end{array}$$

To check your answer, put 9 in place of x in the original problem. Does $5 + 9 = 14$? Yes, it does.

Example 2:
$$\begin{array}{r} y - 16 = 27 \\ +16 \quad +16 \\ \hline y = 43 \end{array}$$

Again, the 16 has to move. To move it to the other side of the equation, we do the **opposite** of **subtract**. We **add** 16 to both sides. Check by putting 43 in place of the y in the original problem. Does $43 - 16 = 27$? Yes.

Solve the problems below.

1. $n + 9 = 27$
2. $12 + y = 55$
3. $51 + v = 67$
4. $f + 16 = 31$
5. $5 + x = 23$

6. $15 + x = 24$
7. $w - 14 = 89$
8. $t - 26 = 20$
9. $m - 12 = 17$
10. $c - 7 = 21$

11. $k - 5 = 29$
12. $a + 17 = 45$
13. $d + 26 = 56$
14. $15 + x = 56$
15. $y + 19 = 32$

16. $t - 16 = 28$
17. $m + 14 = 37$
18. $y - 21 = 29$
19. $f + 7 = 31$
20. $h - 12 = 18$

21. $r - 12 = 37$
22. $h - 17 = 22$
23. $x - 37 = 46$
24. $r - 11 = 28$
25. $t - 5 = 52$

2.2   One-Step Algebra Problems with Multiplication and Division

Solving one-step algebra problems with multiplication and division are just as easy as adding and subtracting. Again, you perform the **opposite** operation. If the problem is a **multiplication** problem, you **divide** to find the answer. If it is a **division** problem, you **multiply** to find the answer. Carefully read the examples below, and you will see how easy they are.

Example 3: $4x = 20$ (4x means 4 times x. 4 is the coefficient of x.)

The goal is to get the numbers on one side of the equal sign and the variable x on the other side. In this problem, the 4 and the x are on the same side of the equal sign. The 4 has to be moved over. $4x$ means 4 times x. The opposite of **multiply** is **divide**. If we divide both sides of the equation by 4, we will find the answer.

$4x = 20$ **We need to divide both sides by 4.**

This means divide by 4. ⟶ $\frac{4x}{4} = \frac{20}{4}$ **We see that $1x = 5$, so $x = 5$.**

When you put 5 in place of _x_ in the original problem, it is correct. $4 \times 5 = 20$

Example 4: $\frac{y}{4} = 2$

This problem means y divided by 4 is equal to 2. In this case, the opposite of divide is multiply. We need to multiply both sides of the equation by 4.

$4 \times \frac{y}{4} = 2 \times 4$ so $y = 8$

When you put 8 in place of _y_ in the original problem, it is correct. $\frac{8}{4} = 2$

Solve the problems below.

1. $2x = 14$

2. $\dfrac{w}{5} = 11$

3. $3h = 45$

4. $\dfrac{x}{4} = 36$

5. $\dfrac{x}{3} = 9$

6. $6d = 66$

7. $\dfrac{w}{9} = 3$

8. $7r = 98$

9. $\dfrac{y}{3} = 2$

10. $10y = 30$

11. $\dfrac{r}{4} = 7$

12. $8t = 96$

13. $\dfrac{z}{2} = 15$

14. $\dfrac{n}{9} = 5$

15. $4z = 24$

16. $6d = 84$

17. $\dfrac{t}{3} = 3$

18. $\dfrac{m}{6} = 9$

19. $9p = 72$

20. $5a = 60$

Sometimes the answer to the algebra problem is a **fraction**. Read the example below, and you will see how easy it is.

Example 5: $4x = 5$

Problems like this are solved just like the problems above and those on the previous page. The only difference is that the answer is a fraction.

In this problem, the 4 is **multiplied** by x. To solve, we need to divide both sides of the equation by 4.

$4x = 5$ Now **divide** by 4. $\dfrac{4x}{4} = \dfrac{5}{4}$ Now cancel. $\dfrac{\cancel{4}x}{\cancel{4}} = \dfrac{5}{4}$ So $x = \dfrac{5}{4}$

When you put $\dfrac{5}{4}$ in place of x in the original problem, it is correct.

$4 \times \dfrac{5}{4} = 5$ Now cancel. $\longrightarrow$ $\cancel{4} \times \dfrac{5}{\cancel{4}} = 5$ So $5 = 5$

Solve the problems below. Some of the answers will be fractions. Some answers will be integers.

1. $2x = 3$

2. $4y = 5$

3. $5t = 2$

4. $12b = 144$

5. $9a = 72$

6. $8y = 16$

7. $7x = 21$

8. $4z = 64$

9. $7x = 126$

10. $6p = 10$

11. $2n = 9$

12. $5x = 11$

13. $15m = 180$

14. $5h = 21$

15. $3y = 8$

16. $2t = 10$

17. $3b = 2$

18. $5c = 14$

19. $4d = 3$

20. $5z = 75$

21. $9y = 4$

22. $7d = 12$

23. $2w = 13$

24. $9g = 81$

25. $6a = 18$

26. $2p = 16$

27. $15w = 3$

28. $5x = 13$

2.3   Multiplying and Dividing with Negative Numbers

Example 6: $-3x = 15$

Step 1:   In the problem, -3 is **multiplied** by x. To find the solution, we must do the opposite. The opposite of **multiply** is **divide**. We must divide both sides of the equation by -3.
$$\frac{-3x}{-3} = \frac{15}{-3}$$

Step 2:   Then cancel.

$$\frac{\cancel{-3}x}{\cancel{-3}} = \frac{-15}{-3} = -5 \qquad x = -5$$

Example 7: $\dfrac{y}{-4} = -20$

Step 1:   In this problem, y is **divided** by -4. To find the answer, do the opposite. **Multiply** both sides by -4.
$$\cancel{-4} \times \frac{y}{\cancel{-4}} = (-20) \times (-4) \qquad \text{so } y = 80$$

Example 8: $-6a = 2$

Step 1:   The answer to an algebra problem can also be a negative fraction.
$$\frac{\cancel{6}a}{\cancel{6}} = \frac{2}{-6} \longleftarrow \text{reduce to get} \quad a = \frac{1}{-3} \quad \text{or} \quad -\frac{1}{3}$$

Note: A negative fraction can be written several different ways.

$$\frac{1}{-3} = \frac{-1}{3} = -\frac{1}{3} = -\left(\frac{1}{3}\right)$$

All mean the same thing

Solve the problems below. Reduce any fractions to lowest terms.

1. $2z = -6$

2. $\dfrac{y}{-5} = 20$

3. $-6k = 54$

4. $4x = -24$

5. $\dfrac{t}{7} = -4$

6. $\dfrac{r}{-2} = -10$

7. $9x = 72$

8. $\dfrac{x}{-6} = 3$

9. $\dfrac{w}{-11} = 5$

10. $5y = -35$

11. $\dfrac{x}{-4} = -9$

12. $7t = -49$

13. $-14x = -28$

14. $\dfrac{m}{3} = -12$

15. $\dfrac{c}{-6} = -6$

16. $\dfrac{d}{8} = -7$

17. $\dfrac{y}{-9} = -4$

18. $-15w = -60$

19. $-12v = 36$

20. $-8z = 32$

21. $-4x = -3$

22. $-12y = 7$

23. $\dfrac{a}{-2} = 22$

24. $-18b = 6$

25. $13a = -36$

26. $\dfrac{b}{-2} = -14$

27. $-24x = -6$

28. $\dfrac{y}{-9} = -6$

29. $\dfrac{x}{-23} = -1$

30. $7x = -7$

31. $-9y = -1$

32. $\dfrac{d}{5} = -10$

33. $\dfrac{z}{-13} = -2$

34. $-5c = 45$

35. $2d = -3$

36. $-8d = -12$

37. $-24w = 9$

38. $-6p = 42$

39. $-9a = -18$

40. $\dfrac{p}{-2} = 15$

2.4 Variables with a Coefficient of Negative One

The answer to an algebra problem should not have a negative sign in front of the variable. For example, the problem $-x = 5$ is not completely solved. Study the examples below to learn how to finish solving this problem.

Example 9: $\quad -x = 5$

$-x$ means the same thing as $-1x$ or -1 times x. To solve this problem, **multiply** both sides by -1.

$(-1)(-1x) = (-1)(5) \quad$ so $\ x = -5$

Example 10: $\quad -y = 3 \quad$ Solve the same way.

$(-1)(-y) = (-1)(-3) \quad$ so $\ y = 3$

Solve the following problems.

1. $-w = 14$

2. $-a = 20$

3. $-x = -15$

4. $-x = -25$

5. $-y = -16$

6. $-t = 62$

7. $-p = -34$

8. $-m = 81$

9. $-w = 17$

10. $-v = -9$

11. $-k = 13$

12. $-q = 7$

2.5   Graphing Inequalities

An inequality is a sentence that contains a $\neq$, $<$, $>$, $\leq$, or $\geq$ sign. Look at the following graphs of inequalities on a number line.

NUMBER LINE

$x < 3$ is read "x is less than 3."

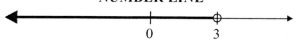

There is no line under the $<$ sign, so the graph uses an **open** endpoint to show x is less than 3 but does not include 3.

$x \leq 5$ is read "x is less than or equal to 5."

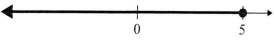

If you see a line under $<$ or $>$ ($\leq$ or $\geq$), the endpoint is filled in. The graph uses a **closed** circle because the number 5 is included in the graph.

$x > -2$ is read "x is greater than -2."

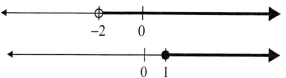

$x \geq 1$ is read "x is greater than or equal to 1."

There can be more than one inequality sign. These are called **compound inequalities**. For example:

$-2 \leq x < 4$ is read "-2 is less than or equal to x and x is less than 4."

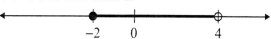

$x < 1$ or $x \geq 4$ is read "x is less than 1 or x is greater than or equal to 4."

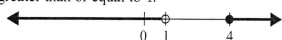

Graph the solution sets of the following inequalities.

1. $x > 8$

2. $x \leq 5$

3. $-5 < x < 1$

4. $x > 7$

5. $1 \leq x < 4$

6. $x < -2$ or $x > 1$

7. $x \geq 10$

8. $x < 4$

9. $x \leq 3$ or $x \geq 5$

10. $x < -1$ or $x > 1$

Give the inequality represented by each of the following number lines.

11.  _____

12. _____

13. _____

14. _____

15. _____

16. _____

17. _____

18.  _____

2.6   Solving Inequalities by Addition and Subtraction

If you add or subtract the same number to both sides of an inequality, the inequality remains the same. It works just like an equation.

Example 11:   Solve and graph the solution set for $x - 2 \leq 5$.

Step 1:   Add 2 to both sides of the inequality.
$$\begin{array}{r} x - 2 \leq 5 \\ +2 \ +2 \\ \hline x \leq 7 \end{array}$$

Step 2:   Graph the solution set for the inequality.

$$0 \qquad\qquad 7$$

Solve and graph the solution set for the following inequalities.

1. $x + 5 > 3$ ⟷

2. $x - 10 < 5$ ⟷

3. $x - 2 \leq 1$ ⟷

4. $9 + x \geq 7$ ⟷

5. $x - 4 > -2$ ⟷

6. $x + 11 \leq 20$ ⟷

7. $x - 3 < -12$ ⟷

8. $x + 6 \geq -3$ ⟷

9. $x + 12 \leq 8$ ⟷

10. $15 + x > 5$ ⟷

11. $x - 6 < -2$ ⟷

12. $x + 7 \geq 4$ ⟷

13. $14 + x \leq 8$ ⟷

14. $x - 8 > 24$ ⟷

15. $x + 1 \leq 12$ ⟷

16. $11 + x \geq 11$ ⟷

17. $x - 3 < 17$ ⟷

18. $x + 9 > -4$ ⟷

19. $x + 6 \leq 14$ ⟷

20. $x - 8 \geq 19$ ⟷

2.7   Solving Inequalities by Multiplication and Division

If you multiply or divide both sides of an inequality by a **positive** number, the inequality symbol stays the same. However, if you multiply or divide both sides of an inequality by a **negative** number, **you must reverse the direction of the inequality symbol.**

Example 12:   Solve and graph the solution set for $4x \leq 20$.

Step 1:   Divide both sides of the inequality by 4. $\dfrac{\overset{1}{\cancel{4}x}}{\underset{1}{\cancel{4}}} \leq \dfrac{\overset{5}{\cancel{20}}}{\underset{1}{\cancel{4}}}$

Step 2:   Graph the solution. $x \leq 5$

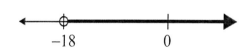

Example 13:   Solve and graph the solution set for $6 > -\dfrac{x}{3}$.

Step 1:   Multiply both sides by -3 and **reverse the direction of the symbol.**

$$(-3) \times 6 < \dfrac{x}{\cancel{-3}} \times \cancel{-3}$$

Step 2:   Graph the solution. $-18 < x$

Solve and graph the following inequalities.

1. $\dfrac{x}{5} > 4$

2. $2x \leq 24$

3. $-6x \geq 36$

4. $\dfrac{x}{10} > -2$

5. $-\dfrac{x}{4} > 8$

6. $-7x \leq -49$

7. $-3x > 18$

8. $-\dfrac{x}{7} \geq 9$

9. $9x \leq 54$

10. $\dfrac{x}{8} > 1$

11. $-\dfrac{x}{9} \leq 3$

12. $-4x < -12$

13. $-\dfrac{x}{2} \geq -20$

14. $10x \leq 30$

15. $\dfrac{x}{12} > -4$

16. $-6x < 24$

Chapter 2 Review

Solve the following one-step algebra problems.

1. $5y = -25$

2. $x + 4 = 24$

3. $d - 11 = 14$

4. $\dfrac{a}{6} = -8$

5. $-t = 2$

6. $-14b = 12$

7. $\dfrac{c}{-10} = -3$

8. $z - 15 = -19$

9. $-13d = 4$

10. $\dfrac{x}{-14} = 2$

11. $-4k = -12$

12. $y + 13 = 27$

13. $15 + h = 4$

14. $14p = 2$

15. $\dfrac{b}{4} = 11$

16. $p - 26 = 12$

17. $x + (-2) = 5$

18. $m + 17 = 27$

19. $\dfrac{k}{-4} = 13$

20. $-18a = -7$

21. $21t = -7$

22. $z - (-9) = 14$

23. $23 + w = 28$

24. $n - 35 = -16$

25. $-a = 26$

26. $-19 + f = -9$

27. $\dfrac{w}{11} = 3$

28. $-7y = 28$

29. $x + 23 = 20$

30. $z - 12 = -7$

31. $-16 + g = 40$

32. $\dfrac{m}{-3} = -9$

33. $d + (-6) = 17$

34. $-p = 47$

35. $k - 16 = 5$

36. $9y = -3$

37. $-2z = -36$

38. $10h = 12$

39. $w - 16 = 4$

40. $y + 10 = -8$

Graph the solution sets of the following inequalities.

41. $x \le -3$

42. $x > 6$

43. $x < -2$

44. $x \ge 4$

Give the inequality represented by each of the following number lines.

45.  _____

46.  _____

47. 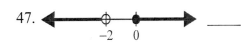 _____

48.  _____

Solve and graph the solution set for the following inequalities.

49. $x - 2 > 8$ ⟵―――――――――⟶   55. $-\dfrac{x}{3} \le 5$ ⟵―――――――――⟶

50. $4 + x < -1$ ⟵―――――――――⟶   56. $x + 10 \le 4$ ⟵―――――――――⟶

51. $6x \ge 54$ ⟵―――――――――⟶   57. $x - 6 \ge -2$ ⟵―――――――――⟶

52. $-2x \le 8$ ⟵―――――――――⟶   58. $7x < -14$ ⟵―――――――――⟶

53. $\dfrac{x}{2} > -1$ ⟵―――――――――⟶   59. $-3x > -12$ ⟵―――――――――⟶

54. $-x < -9$ ⟵―――――――――⟶   60. $-\dfrac{x}{6} \le -3$ ⟵―――――――――⟶

Chapter 2 Test

1. What is the value n?
   $n + 9 = 27$

   A 3
   B 36
   C 18
   D 19

2. What is the value of y?
   $y - 21 = 29$

   A 50
   B 8
   C 49
   D 54

3. Solve for x: $4x = 20$

   A $\dfrac{20}{4}$
   B 5
   C $\dfrac{10}{2}$
   D $\dfrac{1}{4}$

4. Solve for x: $\dfrac{x}{4} = 16$

   A 4
   B 24
   C 32
   D 64

5. What is $-6x \geq 36$ in simplest terms?

   A $x \geq -6$
   B $x \leq -6$
   C $x \geq -\dfrac{1}{6}$
   D $x \leq -\dfrac{1}{6}$

6. Solve for x: $-3x = 15$

   A 5
   B 3
   C -5
   D $-\dfrac{1}{5}$

7. Solve for x: $\dfrac{x}{-6} = -6$

   A 36
   B -36
   C 1
   D -1

8. Which inequality does the number line represent?

   A $x \leq -1$
   B $x < -1$
   C $x > -1$
   D $x \geq -1$

9. Which inequality does the number line represent?

   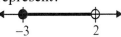

   A $x \geq 3$
   B $x \geq 3, x \leq 2$
   C $-3 \leq x < 2$
   D $-3 < x \leq 2$

10. What is $x - 2 \leq 5$ in simplest terms?

    A $x \leq 7$
    B $x \leq 9$
    C $x \leq 5$
    D $x \leq 4$

Chapter 3
Solving Multi-Step Equations and Inequalities

This chapter covers the following Alabama objectives and standards in mathematics:

HSGE

Mathematics

	Objective(s)
Standard I	1, 2
Standard II	1, 4
Standard V	3

3.1 Two-Step Algebra Problems

In the following two-step algebra problems, **additions** and **subtractions** are performed first and then **multiplication** and **division**.

Example 1: $-4x + 7 = 31$

Step 1: Subtract 7 from both sides.

$$\begin{array}{rr} -4x + 7 & = 31 \\ -7 & -7 \\ \hline -4x & = 24 \end{array}$$

Step 2: Divide both sides by -4.

$$\frac{-4x}{-4} = \frac{24}{-4} \qquad \text{so } x = -6$$

Example 2: $-8 - y = 12$

Step 1: Add 8 to both sides.

$$\begin{array}{rr} -8 - y & = 12 \\ +8 & +8 \\ \hline -y & = 20 \end{array}$$

Step 2: To finish solving a problem with a negative sign in front of the variable, multiply both sides by -1. The variable needs to be positive in the answer.

$$(-1)(-y) = (-1)(20) \text{ so } y = -20$$

Solve the two-step algebra problems below.

1. $6x - 4 = -34$

2. $5y - 3 = 32$

3. $8 - t = 1$

4. $10p - 6 = -36$

5. $11 - 9m = -70$

6. $4x - 12 = 24$

7. $3x - 17 = -41$

8. $9d - 5 = 49$

9. $10h + 8 = 78$

10. $-6b - 8 = 10$

11. $-g - 24 = -17$

12. $-7k - 12 = 30$

13. $9 - 5r = 64$

14. $6y - 14 = 34$

15. $12f + 15 = 51$

16. $21t + 17 = 80$

17. $20y + 9 = 149$

18. $15p - 27 = 33$

19. $22h + 9 = 97$

20. $-5 + 36w = 175$

3.2   Two-Step Algebra Problems with Fractions

An algebra problem may contain a fraction. Study the following example to understand how to solve algebra problems that contain a fraction.

Example 3: $\dfrac{x}{2} + 4 = 3$

Step 1:
$$\dfrac{x}{2} + 4 = 3$$
$$\underline{\quad -4 \qquad -4 \quad}$$
$$\dfrac{x}{2} \qquad = -1$$
Subtract 4 from both sides.

Step 2: $\dfrac{x}{2} = -1$ Multiply both sides by 2 to eliminate the fraction.

$$\dfrac{x}{\cancel{2}} \times \cancel{2} = -1 \times 2, \; x = -2$$

Simplify the following algebra problems.

1. $4 + \dfrac{y}{3} = 7$

2. $\dfrac{a}{2} + 5 = 12$

3. $\dfrac{w}{5} - 3 = 6$

4. $\dfrac{x}{9} - 9 = -5$

5. $\dfrac{b}{6} + 2 = -4$

6. $7 + \dfrac{z}{2} = -13$

7. $\dfrac{x}{2} - 7 = 3$

8. $\dfrac{c}{5} + 6 = -2$

9. $3 + \dfrac{x}{11} = 7$

10. $16 + \dfrac{m}{6} = 14$

11. $\dfrac{p}{3} + 5 = -2$

12. $\dfrac{t}{8} + 9 = 3$

13. $\dfrac{v}{7} - 8 = -1$

14. $5 + \dfrac{h}{10} = 8$

15. $\dfrac{k}{7} - 9 = 1$

16. $\dfrac{y}{4} + 13 = 8$

17. $15 + \dfrac{z}{14} = 13$

18. $\dfrac{b}{6} - 9 = -14$

19. $\dfrac{d}{3} + 7 = 12$

20. $10 + \dfrac{b}{6} = 4$

21. $2 + \dfrac{p}{4} = -6$

22. $\dfrac{t}{7} - 9 = -5$

23. $\dfrac{a}{10} - 1 = 3$

24. $\dfrac{a}{8} + 16 = 9$

3.3   More Two-Step Algebra Problems with Fractions

Study the following example to understand how to solve algebra problems that contain a different type of fraction.

Example 4: $\dfrac{x+2}{4} = 3$ In this example, "$x + 2$" is divided by 4, and not just the x or the 2.

Step 1: $\dfrac{x+2}{4} \times 4 = 3 \times 4$ First multiply both sides by 4 to eliminate the fraction.

Step 2: $\begin{array}{r} x + 2 = 12 \\ -2 \quad\; -2 \\ \hline x \;\;= 10 \end{array}$ Next, subtract 2 from both sides.

Solve the following problems.

1. $\dfrac{x+1}{5} = 4$

2. $\dfrac{z-9}{2} = 7$

3. $\dfrac{b-4}{4} = -5$

4. $\dfrac{y-9}{3} = 7$

5. $\dfrac{d-10}{-2} = 12$

6. $\dfrac{w-10}{-8} = -4$

7. $\dfrac{x-1}{-2} = -5$

8. $\dfrac{c+40}{-5} = -7$

9. $\dfrac{13+h}{2} = 12$

10. $\dfrac{k-10}{3} = 9$

11. $\dfrac{a+11}{-4} = 4$

12. $\dfrac{x-20}{7} = 6$

13. $\dfrac{t+2}{6} = -5$

14. $\dfrac{b+1}{-7} = 2$

15. $\dfrac{f-9}{3} = 8$

16. $\dfrac{4+w}{6} = -6$

17. $\dfrac{3+t}{3} = 10$

18. $\dfrac{x+5}{5} = -3$

19. $\dfrac{g+3}{2} = 11$

20. $\dfrac{k+1}{-6} = 5$

21. $\dfrac{y-14}{2} = -8$

22. $\dfrac{z-4}{-2} = 13$

23. $\dfrac{w+2}{15} = -1$

24. $\dfrac{3+h}{3} = 6$

3.4   Combining Like Terms

In algebra problems, separate **terms** by $+$ and $-$ signs. The expression $5x - 4 - 3x + 7$ has 4 terms: $5x$, 4, $3x$, and 7. Terms having the same variable can be combined (added or subtracted) to simplify the expression. $5x - 4 - 3x + 7$ simplifies to $2x + 3$.

$$5x - 3x \quad - 4 + 7 \ = 2x + 3$$

Simplify the following expressions.

1. $7x + 12x$

2. $8y - 5y + 8$

3. $4 - 2x + 9$

4. $11a - 16 - a$

5. $9w + 3w + 3$

6. $-5x + x + 2x$

7. $w - 15 + 9w$

8. $21 - 10t + 9 - 2t$

9. $-3 + x - 4x + 9$

10. $7b + 12 + 4b$

11. $4h - h + 2 - 5$

12. $-6k + 10 - 4k$

13. $2a + 12a - 5 + a$

14. $5 + 9c - 10$

15. $-d + 1 + 2d - 4$

16. $-8 + 4h + 1 - h$

17. $12x - 4x + 7$

18. $10 + 3z + z - 5$

19. $14 + 3y - y - 2$

20. $11p - 4 + p$

21. $11m + 2 - m + 1$

3.5   Solving Equations with Like Terms

When an equation has two or more like terms on the same side of the equation, combine like terms as the **first** step in solving the equation.

Example 5: $7x + 2x - 7 = 21 + 8$

Step 1:   Combine like terms on both sides of the equation.

Step 2:   Solve the two-step algebra problem as explained previously.

$$\begin{aligned} 7x + 2x - 7 &= 21 + 8 \\ 9x - 7 &= 29 \\ +7 \quad &\quad +7 \\ 9x \div 9 &= 36 \div 9 \\ x &= 4 \end{aligned}$$

Solve the equations below combining like terms first.

1. $3w - 2w + 4 = 6$

2. $7x + 3 + x = 16 + 3$

3. $5 - 6y + 9y = -15 + 5$

4. $-14 + 7a + 2a = -5$

5. $-2t + 4t - 7 = 9$

6. $9d + d - 3d = 14$

7. $-6c - 4 - 5c = 10 + 8$

8. $15m - 9 - 6m = 9$

9. $-4 - 3x - x = -16$

10. $9 - 12p + 5p = 14 + 2$

11. $10y + 4 - 7y = -17$

12. $-8a - 15 - 4a = 9$

If the equation has like terms on both sides of the equation, you must get all of the terms with a **variable** on one side of the equation and all of the **integers** on the other side of the equation.

Example 6: $3x + 2 = 6x - 1$

Step 1:	Subtract $6x$ from both sides to move all the **variables** to the left side.
Step 2:	Subtract 2 from both sides to move all the **integers** to the right side.
Step 3:	Divide by -3 to solve for x.

$$
\begin{aligned}
3x + 2 &= 6x - 1 \\
-6x &\quad -6x \\
-3x + 2 &= -1 \\
-2 &\quad -2 \\
\hline
\frac{-3x}{-3} &= \frac{-3}{-3} \\
x &= 1
\end{aligned}
$$

Solve the following problems.

1. $3a + 1 = a + 9$

2. $2d - 12 = d + 3$

3. $5x + 6 = 14 - 3x$

4. $15 - 4y = 2y - 3$

5. $9w - 7 = 12w - 13$

6. $10b + 19 = 4b - 5$

7. $-7m + 9 = 29 - 2m$

8. $5x - 26 = 13x - 2$

9. $19 - p = 3p - 9$

10. $-7p - 14 = -2p + 11$

11. $16y + 12 = 9y + 33$

12. $13 - 11w = 3 - w$

13. $-17b + 23 = -4 - 8b$

14. $k + 5 = 20 - 2k$

15. $12 + m = 4m + 21$

16. $7p - 30 = p + 6$

17. $19 - 13z = 9 - 12z$

18. $8y - 2 = 4y + 22$

19. $5 + 16w = 6w - 45$

20. $-27 - 7x = 2x + 18$

21. $-12x + 14 = 8x - 46$

22. $27 - 11h = 5 - 9h$

23. $5t + 36 = -6 - 2t$

24. $17y + 42 = 10y + 7$

25. $22x - 24 = 14x - 8$

26. $p - 1 = 4p + 17$

27. $4d + 14 = 3d - 1$

28. $7w - 5 = 8w + 12$

29. $-3y - 2 = 9y + 22$

30. $17 - 9m = m - 23$

3.6   Removing Parentheses

The distributive principle is used to remove parentheses.

Example 7: $2(a + 6)$

You multiply 2 by each term inside the parentheses. $2 \times a = 2a$ and $2 \times 6 = 12$. The 12 is a positive number so use a plus sign between the terms in the answer.
$2(a + 6) = 2a + 12$

Example 8: $4(-5c + 2)$

The first term inside the parentheses can be negative. Multiply in exactly the same way as the examples above. $4 \times (-5c) = -20c$ and $4 \times 2 = 8$
$4(-5c + 2) = -20c + 8$

Remove the parentheses in the problems below.

1. $7(n + 6)$
2. $8(2g - 5)$
3. $11(5z - 2)$
4. $6(-y - 4)$
5. $3(-3k + 5)$

6. $4(d - 8)$
7. $2(-4x + 6)$
8. $7(4 + 6p)$
9. $5(-4w - 8)$
10. $6(11x + 2)$

11. $10(9 - y)$
12. $9(c - 9)$
13. $12(-3t + 1)$
14. $3(4y + 9)$
15. $8(b + 3)$

The number in front of the parentheses can also be negative. Remove these parentheses the same way.

Example 9: $-2(b - 4)$

First, multiply $-2 \times b = -2b$
Second, multiply $-2 \times -4 = 8$
Copy the two products. The second product is a positive number so put a plus sign between the terms in the answer.
$-2(b - 4) = -2b + 8$

Remove the parentheses in the following problems.

16. $-7(x + 2)$
17. $-5(4 - y)$
18. $-4(2b - 2)$
19. $-2(8c + 6)$
20. $-5(-w - 8)$

21. $-3(4x - 2)$
22. $-2(-z + 2)$
23. $-4(7p + 7)$
24. $-9(t - 6)$
25. $-10(2w + 4)$

26. $-3(9 - 7p)$
27. $-9(-k - 3)$
28. $-1(7b - 9)$
29. $-6(-5t - 2)$
30. $-7(-v + 4)$

3.7   Multi-Step Algebra Problems

You can now use what you know about removing parentheses, combining like terms, and solving simple algebra problems to solve problems that involve three or more steps. Study the examples below to see how easy it is to solve multi-step problems.

Example 10: $3(x+6) = 5x - 2$

Step 1:	Use the distributive property to remove parentheses.	$3x + 18 = 5x - 2$
Step 2:	Subtract $5x$ from each side to move the terms with variables to the left side of the equation.	$\dfrac{-5x \qquad -5x}{-2x + 18 = -2}$
Step 3:	Subtract 18 from each side to move the integers to the right side of the equation.	$\dfrac{-18 \quad -18}{}$
Step 4:	Divide both sides by -2 to solve for x.	$\dfrac{-2x}{-2} = \dfrac{-20}{-2}$
		$x = 10$

Example 11: $\dfrac{3(x-3)}{2} = 9$

Step 1:	Use the distributive property to remove parentheses.	$\dfrac{3x - 9}{2} = 9$
Step 2:	Multiply both sides by 2 to eliminate the fraction.	$\dfrac{2(3x-9)}{2} = 2(9)$
Step 3:	Add 9 to both sides, and combine like terms.	$3x - 9 = 18$
		$\quad +9 \qquad +9$
Step 4:	Divide both sides by 3 to solve for x.	$\dfrac{3x}{3} = \dfrac{27}{3}$
		$x = 9$

Solve the following multi-step algebra problems.

1. $2(y-3) = 4y + 6$

2. $\dfrac{2(a+4)}{2} = 12$

3. $\dfrac{10(x-2)}{5} = 14$

4. $\dfrac{12y - 18}{6} = 4y + 3$

5. $2x + 3x = 30 - x$

6. $\dfrac{2a+1}{3} = a + 5$

7. $5(b-4) = 10b + 5$

8. $-8(y+4) = 10y + 4$

9. $\dfrac{x+4}{-3} = 6 - x$

10. $\dfrac{4(n+3)}{5} = n - 3$

11. $3(2x - 5) = 8x - 9$

12. $7 - 10a = 9 - 9a$

13. $7 - 5x = 10 - (6x + 7)$

14. $4(x - 3) - x = x - 6$

15. $4a + 4 = 3a - 4$

16. $-3(x - 4) + 5 = -2x - 2$

17. $5b - 11 = 13 - b$

18. $\dfrac{-4x + 3}{2x} = \dfrac{7}{2x}$

19. $-(x + 1) = -2(5 - x)$

20. $4(2c + 3) - 7 = 13$

21. $6 - 3a = 9 - 2(2a + 5)$

22. $-5x + 9 = -3x + 11$

23. $3y + 2 - 2y - 5 = 4y + 3$

24. $3y - 10 = 4 - 4y$

25. $-(a + 3) = -2(2a + 1) - 7$

26. $5m - 2(m + 1) = m - 10$

27. $\dfrac{1}{2}(b - 2) = 5$

28. $-3(b - 4) = -2b$

29. $4x + 12 = -2(x + 3)$

30. $\dfrac{7x + 4}{3} = 2x - 1$

31. $9x - 5 = 8x - 7$

32. $7x - 5 = 4x + 10$

33. $\dfrac{4x + 8}{2} = 6$

34. $2(c + 4) + 8 = 10$

35. $y - (y + 3) = y + 6$

36. $4 + x - 2(x - 6) = 8$

3.8   Multi-Step Inequalities

Remember that adding and subtracting with inequalities follow the same rules as equations. When you multiply or divide both sides of an inequality by the same positive number, the rules are also the same as for equations. However, when you multiply or divide both sides of an inequality by a **negative** number, you must **reverse** the inequality symbol.

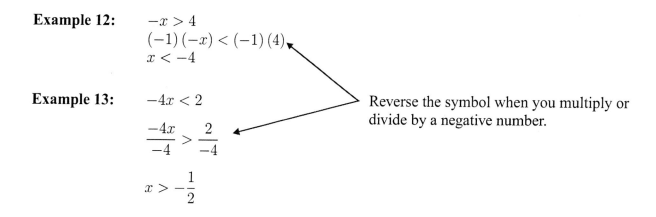

Example 12: $-x > 4$
$(-1)(-x) < (-1)(4)$
$x < -4$

Example 13: $-4x < 2$

$\dfrac{-4x}{-4} > \dfrac{2}{-4}$

$x > -\dfrac{1}{2}$

Reverse the symbol when you multiply or divide by a negative number.

When solving multi-step inequalities, first add and subtract to isolate the term with the variable. Then multiply and divide.

Example 14: $2x - 8 > 4x + 1$

Step 1:   Add 8 to both sides.

$2x - 8 + 8 > 4x + 1 + 8$
$2x > 4x + 9$

Step 2:   Subtract $4x$ from both sides.

$2x - 4x > 4x + 9 - 4x$
$-2x > 9$

Step 3:   Divide by -2. Remember to change the direction of the inequality sign.

$\dfrac{-2x}{-2} < \dfrac{9}{-2}$

$x < -\dfrac{9}{2}$

Solve each of the following inequalities.

1. $8 - 3x \leq 7x - 2$

2. $3(2x - 5) \geq 8x - 5$

3. $\frac{1}{3}b - 2 > 5$

4. $7 + 3y > 2y - 5$

5. $3a + 5 < 2a - 6$

6. $3(a - 2) > -5a - 2(3 - a)$

7. $2x - 7 \geq 4(x - 3) + 3x$

8. $6x - 2 \leq 5x + 5$

9. $-\frac{x}{4} > 12$

10. $-\frac{2x}{3} \leq 6$

11. $3b + 5 < 2b - 8$

12. $4x - 5 \leq 7x + 13$

13. $4x + 5 \leq -2$

14. $2y - 5 > 7$

15. $4 + 2(3 - 2y) \leq 6y - 20$

16. $-4c + 6 \leq 8$

17. $-\frac{1}{2}x + 2 > 9$

18. $\frac{1}{4}y - 3 \leq 1$

19. $-3x + 4 > 5$

20. $\frac{y}{2} - 2 \geq 10$

21. $7 + 4c < -2$

22. $2 - \frac{a}{2} > 1$

23. $10 + 4b \leq -2$

24. $-\frac{1}{2}x + 3 > 4$

3.9   Compound Linear Inequalities

Inequalities can have more than one inequality sign. These are called **compound inequalities**. For example:

$$-2 \leq x < 4 \text{ is read "}-2 \text{ is less than or equal to } x \text{ and } x \text{ is less than 4."}$$

You learned earlier in this chapter that you can perform any operation on an equation, as long as you perform the same operation to the other side. The same is true of inequalities. It's also true of compound inequalities.

Example 15:   Solve the combined inequality $-20 < 5x - 10 < 20$ for x. Graph the solution set on a number line.

Step 1:   We must isolate x. The first step is to add 10 to each part of the combined inequality.

$$
\begin{array}{ccccc}
-20 & < & 5x - 10 & < & 20 \\
+10 & & +10 & & +10 \\
\hline
-10 & < & 5x & < & 30
\end{array}
$$

Step 2:   Now divide each part by 5.

$$\frac{-10}{5} < \frac{5x}{5} < \frac{30}{5}$$

$$-2 < x < 6$$

Solve the inequalities and graph the solution set on a number line.

1. $-14 < 2x + 4 \leq 14$

2. $6 > 4b - 8 > 0$

3. $-36 < 12e + 3 < 36$

4. $49 > 8f - 15 > 1$

5. $-98 < 7g - 91 < 0$

6. $14 > 4a + 10 \geq -14$

7. $13 \leq 9k - 32 < 121$

8. $35 > -5j - 50 > -10$

9. $43 > 6f - 83 > 19$

10. $43 > 31 + 6x \geq 4$

11. $7 > 9d - 8 > -10$

12. $48 > 10c - 12 > 23$

13. $-6 < 29 + 7v < 15$

14. $-2 < 9e + 7 \leq 11$

15. $77 \geq 8a + 21 > 29$

16. $-4 \leq 3g - 40 \leq 38$

17. $2 < 10 - 8r < 34$

18. $37 \geq 11u - 29 \geq 4$

Chapter 3 Review

Solve each of the following equations.

1. $4a - 8 = 28$

2. $5 + \dfrac{x}{8} = -4$

3. $-7 + 23w = 108$

4. $\dfrac{y - 8}{6} = 7$

5. $c - 13 = 5$

6. $\dfrac{b + 9}{12} = -3$

Simplify the following expressions by combining like terms.

7. $-4a + 8 + 3a - 9$

8. $14 + 2z - 8 - 5z$

9. $-7 - 7x - 2 - 9x$

Solve.

10. $19 - 8d = d - 17$

11. $7w - 8w = -4w - 30$

12. $6 + 16x = -2x - 12$

13. $6(b - 4) = 8b - 18$

14. $4x - 16 = 7x + 2$

15. $9w - 2 = -w - 22$

Remove parentheses.

16. $3(-4x + 7)$

17. $11(2y + 5)$

18. $6(8 - 9b)$

19. $-8(-2 + 3a)$

20. $-2(5c - 3)$

21. $-5(7y - 1)$

Solve each of the following equations and inequalities.

22. $\dfrac{-11c - 35}{4} = 4c - 2$

23. $5 + x - 3(x + 4) = -17$

24. $4(2x + 3) \geq 2x$

25. $7 - 3x \leq 6x - 2$

26. $\dfrac{5(n + 4)}{3} = n - 8$

27. $-y > 14$

28. $2(3x - 1) \geq 3x - 7$

29. $3(x + 2) < 7x - 10$

Solve each of the following compound inequalities and graph the solution set on a number line.

30. $-22 < -11k < 22$

31. $18 > -3w > -18$

32. $21 \leq 5x + 6 \leq 46$

33. $20 > 3x + 2 \geq 1$

Chapter 3 Test

1. Find the value of n. $19n - 57 = 76$

   A 1
   B 3
   C 5
   D 7

2. Solve for x. $14x + 84 = 154$

   A 4
   B 5
   C 11
   D 17

3. Which of the following is equivalent to $4 - 5x > 3(x - 4)$?

   A $4 - 5x > 3x - 4$
   B $4 - 5x > 3x - 12$
   C $4 - 5x > 3x - 1$
   D $4 - 5x > 3x - 7$

4. Which of the following is equivalent to $3(x - 2) + 1 - 2x = -4$?

   A $x - 6 = -4$
   B $-6x + 1 = -4$
   C $5x - 7 = -4$
   D $x - 5 = -4$

5. $5(2x + 11) - 3 \times 5 = ?$

   A $7x + 40$
   B $7x + 20$
   C $10x + 40$
   D $10x + 260$

6. Solve: $4b - 8 < 56$

   A $b < 12$
   B $b < 16$
   C $b < -12$
   D $b < -16$

7. Solve: $3(x - 2) - 1 = 6(x + 5)$

   A -4
   B $-\dfrac{37}{3}$
   C 4
   D $\dfrac{23}{3}$

8. Which of the following is equivalent to $3(2x - 5) - 4(x - 3) = 7$?

   A $x + 27 = 7$
   B $2x - 3 = 7$
   C $10x - 27 = 7$
   D $x - 27 = 7$

9. Which of these graphs represents the solution of $36 \geq -6x - 18 > 18$?

10. $-x + 2 < 10$ is true for

   A All values of x.
   B No values of x.
   C $x > -8$.
   D $x < 8$.

Chapter 4
Ratios and Proportions

This chapter covers the following Alabama objectives and standards in mathematics:

	Objective(s)
Standard VII	7

4.1 Ratio Problems

In some word problems, you may be asked to express answers as a **ratio**. Ratios can look like fractions. Numbers must be written in the order they are requested. In the following problem, 8 cups of sugar is mentioned before 6 cups of strawberries. But in the question part of the problem, you are asked for the ratio of STRAWBERRIES to SUGAR. The amount of strawberries IS THE FIRST WORD MENTIONED, so it must be the **top** number of the fraction. The amount of sugar, THE SECOND WORD MENTIONED, must be the **bottom** number of the fraction.

Example 1: The recipe for jam requires 8 cups of sugar for every 6 cups of strawberries. What is the ratio of strawberries to sugar in this recipe?

First number requested $\underline{6}$ cups strawberries

Second number requested   8   cups sugar

Answers may be reduced to lowest terms. $\frac{6}{8} = \frac{3}{4}$

Practice writing ratios for the following word problems and reduce to lowest terms. DO NOT CHANGE ANSWERS TO MIXED NUMBERS. Ratios should be left in fraction form.

1. Out of the 248 seniors, 112 are boys. What is the ratio of boys to the total number of seniors?

2. It takes 7 cups of flour to make 2 loaves of bread. What is the ratio of cups of flour to loaves of bread?

3. A skyscraper that stands 620 feet tall casts a shadow that is 125 feet long. What is the ratio of the shadow to the height of the skyscraper?

4. Twenty boxes of paper weigh 520 pounds. What is the ratio of boxes to pounds?

5. The newborn weighs 8 pounds and is 22 inches long. What is the ratio of weight to length?

6. Jack paid $6.00 for 10 pounds of apples. What is the ratio of the price of apples to the pounds of apples?

7. Jordan spends $45 on groceries. Of that total, $23 is for steaks. What is the ratio of steak cost to the total grocery cost?

8. Madison's flower garden measures 8 feet long by 6 feet wide. What is the ratio of length to width?

4.2 Solving Proportions

Two **ratios (fractions)** that are **equal** to each other are called **proportions. For example,** $\frac{1}{4} = \frac{2}{8}$. **Read the following example to see how to find a number missing from a proportion.**

Example 2: $\frac{5}{15} = \frac{8}{x}$

Step 1: To find x, you first multiply the two numbers that are diagonal to each other.

$$\frac{5}{\{15\}} = \frac{\{8\}}{x}$$

$15 \times 8 = 120$

$5 \times x = 5x$

Therefore, $5x = 120$

Step 2: Then divide the product (120) by the other number in the proportion (5).

$120 \div 5 = 24$

Therefore, $\frac{5}{15} = \frac{8}{24}$ **and** $x = 24$.

Practice finding the number missing from the following proportions. First, multiply the two numbers that are diagonal from each other. Then divide by the other number.

1. $\frac{2}{5} = \frac{6}{x}$

2. $\frac{9}{3} = \frac{x}{5}$

3. $\frac{x}{12} = \frac{3}{4}$

4. $\frac{7}{x} = \frac{3}{9}$

5. $\frac{12}{x} = \frac{2}{5}$

6. $\frac{12}{x} = \frac{4}{3}$

7. $\frac{27}{3} = \frac{x}{2}$

8. $\frac{1}{x} = \frac{3}{12}$

9. $\frac{15}{2} = \frac{x}{4}$

10. $\frac{7}{14} = \frac{x}{6}$

11. $\frac{5}{6} = \frac{10}{x}$

12. $\frac{4}{x} = \frac{3}{6}$

13. $\frac{x}{5} = \frac{9}{15}$

14. $\frac{9}{18} = \frac{x}{2}$

15. $\frac{5}{7} = \frac{35}{x}$

16. $\frac{x}{2} = \frac{8}{4}$

17. $\frac{15}{20} = \frac{x}{8}$

18. $\frac{x}{40} = \frac{5}{100}$

4.3   Ratio and Proportion Word Problems

Example 3:   A stick one meter long is held perpendicular to the ground and casts a shadow 0.4 meters long. At the same time, an electrical tower casts a shadow 112 meters long. Use ratio and proportion to find the height of the tower.

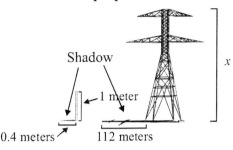

Step 1:   Set up a proportion using the numbers in the problem. Put the shadow lengths on one side of the equation and put the heights on the other side. The 1 meter height is paired with the 0.4 meter length, so let them both be top numbers. Let the unknown height be x.

$$\underset{\text{length}}{\text{shadow}} \qquad \underset{\text{height}}{\text{object}}$$

$$\frac{0.4}{112} = \frac{1}{x}$$

Step 2:   Solve the proportion as you did on page 56.

$$112 \times 1 = 112 \qquad 112 \div 0.4 = 280$$

Answer: The tower height is 280 meters.

Use ratio and proportion to solve the following problems.

1. Rudolph can mow a lawn that measures 1000 square feet in 2 hours. At that rate, how long would it take him to mow a lawn 3500 square feet?

2. Faye wants to know how tall her school building is. On a sunny day, she measures the shadow of the building to be 6 feet. At the same time she measures the shadow cast by a 5 foot statue to be 2 feet. How tall is her school building?

3. Out of every 5 students surveyed, 2 listen to country music. At that rate, how many students in a school of 800 listen to country music?

4. Butterfly, a Labrador retriever, has a litter of 8 puppies. Four are black. At that rate, how many puppies in a litter of 10 would be black?

5. According to the instructions on a bag of fertilizer, 5 pounds of fertilizer are needed for every 100 square feet of lawn. How many square feet will a 25-pound bag cover?

6. A race car can travel 2 laps in 5 minutes. At this rate, how long will it take the race car to complete 100 laps?

7. If it takes 7 cups of flour to make 4 loaves of bread, how many loaves of bread can you make from 35 cups of flour?

8. If 3 pounds of jelly beans cost $6.30, how much would 2 pounds cost?

9. For the first 4 home football games, the concession stand sold a total of 600 hotdogs. If that ratio stays constant, how many hotdogs will sell for all 10 home games?

4.4   Direct Variation

Direct variation occurs in a function when y varies directly, or in the same way, as x varies. The two values vary by a proportional factor, k. The variation is treated just like a proportion.

Example 4:   If y varies directly with x, and $y = 18$ when $x = 12$, what is the value of y when $x = 6$?

Step 1:   Set up the values in a proportion like you did in the previous two sections. Be sure to put the correct corresponding x and y values on the same line within the fractions.

$$\underset{\dfrac{12}{6}}{\underline{x \text{ values}}} = \underset{\dfrac{18}{y}}{\underline{y \text{ values}}}$$

Step 2:   Solve for y by multiplying the diagonals together and setting them equal to one another.

$$12 \times y = 6 \times 18$$

$$\frac{12y}{12} = \frac{108}{12} \qquad \text{Divide both sides by 12.}$$

$$y = 9$$

Solve these direct variation problems.

1. If $y = 6$ and $x = 3$, what is the value of y when $x = 5$?

2. If $y = 10$ and $x = 5$, what is the value of y when $x = 4$?

3. If $y = 6$ and $x = 2$, what is the value of y when $x = 7$?

4. If $y = 8$ and $x = 4$, what is the value of y when $x = 6$?

5. If $y = 15$ and $x = 3$, what is the value of y when $x = 5$?

6. At the local grocery store, 2 pineapples cost $2.78. How much do 5 pineapples cost?

7. It normally takes Jessica 45 minutes to get to her friend's house 20 miles away. Tomorrow she is meeting her friend at the mall, which is 28 miles away from her house. If she travels at the same rate she normally travels, how long will it take her to get to the mall?

4.5   Maps and Scale Drawings

Example 5:   On a map drawn to scale, 5 cm represents 30 kilometers. A line segment connecting two cities is 7 cm long. What distance does this line segment represent?

Step 1:   Set up a proportion using the numbers in the problem. Keep centimeters on one side of the equation and kilometers on the other. The 5 cm is paired with the 30 kilometers, so let them both be top numbers. Let the unknown distance be x.

$$\frac{\text{cm}}{}\quad\frac{\text{km}}{}$$
$$\frac{5}{7} = \frac{30}{x}$$

Step 2:   Solve the proportion as you have previously.

$7 \times 30 = 210$

$210 \div 5 = 42$

Answer: 7 cm represents 42 km.

Sometimes the answer to a scale drawing problem will be a fraction or a mixed number.

Example 6:   On a scale drawing, 2 inches represents 30 feet. How many inches long is a line segment that represents 5 feet?

Step 1:   Set up the proportion as you did above.

$$\frac{\text{inches}}{}\quad\frac{\text{feet}}{}$$
$$\frac{2}{x} = \frac{30}{5}$$

Step 2:   First multiply the two numbers that are diagonal from each other. Then divide by the other number.

$2 \times 5 = 10$

$10 \div 30$ is less than 1 so express the answer as a fraction and reduce.

$10 \div 30 = \dfrac{10}{30} = \dfrac{1}{3}$ inch

Set up proportions for each of the following problems and solve.

1. If 2 inches represents 50 miles on a scale drawing, how long would a line segment be that represents 25 miles?

2. On a scale drawing, 2 cm represents 15 km. A line segment on the drawing is 3 cm long. What distance does this line segment represent?

3. On a map drawn to scale, 5 cm represents 250 km. How many kilometers are represented by a line 6 cm long?

4. If 2 inches represents 80 miles on a scale drawing, how long would a line segment be that represents 280 miles?

5. On a map drawn to scale, 5 cm represents 200 km. How long would a line segment be that represents 260 km?

6. On a scale drawing of a house plan, one inch represents 5 feet. How many feet wide is the bathroom if the width on the drawing is 3 inches?

Chapter 4 Review

Solve the following proportions and ratios.

1. $\dfrac{8}{x} = \dfrac{1}{2}$

2. $\dfrac{2}{5} = \dfrac{x}{10}$

3. $\dfrac{x}{6} = \dfrac{3}{9}$

4. $\dfrac{4}{9} = \dfrac{8}{x}$

5. Out of 100 coins, 45 are in mint condition. What is the ratio of mint condition coins to the total number of coins?

6. The ratio of boys to girls in the ninth grade is 6 : 5. If there are 135 girls in the class, how many boys are there?

7. Twenty out of the total 235 seniors graduate with honors. What is the ratio of seniors graduating with honors to the total number of seniors?

8. Aunt Bess uses 3 cups of oatmeal to bake 6-dozen oatmeal cookies. How many cups of oatmeal would she need to bake 15-dozen cookies?

9. On a map, 2 centimeters represents 150 kilometers. If a line between two cities measures 5 centimeters, how many kilometers apart are they?

10. When Rick measures the shadow of a yard stick, it is 5 inches. At the same time, the shadow of the tree he would like to chop down is 45 inches. How tall is the tree in yards?

11. If 4 inches represents 8 feet on a scale drawing, how many feet does 6 inches represent?

12. On a map scale, 2 centimeters represents 5 kilometers. If two towns on the map are 20 kilometers apart, how long would the line segment be between the two towns on the map?

13. If $y = 10$ and $x = 5$, what is the value of y when $x = 4$? Use direct variation to solve.

14. If $y = 42$ and $x = 6$, what is the value of y when $x = 12$? Use direct variation to solve.

15. Shanice waters 30 plants in 10 minutes. At this same rate, how many plants can she water in 25 minutes?

Chapter 4 Test

1. Solve for x: $\dfrac{3}{4} = \dfrac{9}{x}$

   A 4
   B 27
   C 12
   D 36

2. Solve for x: $\dfrac{3}{x} = \dfrac{9}{27}$

   A 3
   B 6
   C 9
   D 12

3. The ratio of girls to boys at summer camp is 5 : 3. If there are 90 girls, how many boys are there?

   A 54
   B 100
   C 106
   D 113

4. April was practicing jumping rope. She counted off 60 jumps in one minute. If she stays at the same rate, how many times will she jump in 9 minutes?

   A 540
   B 54
   C 600
   D 60

5. It takes Emilio 2 minutes to inflate a basketball using a mechanical pump at the sporting goods store he works at. Emilio has to inflate 75 basketballs for a special sale. If Emilio continues at the same rate, how long will it take him to inflate all 75 basketballs?

   A 75 minutes
   B 73 minutes
   C 148 minutes
   D 150 minutes

6. Solve for x: $\dfrac{x}{4} = \dfrac{7}{28}$

   A 4
   B 3
   C 2
   D 1

7. The scale on a map of the ground floor of a museum is $\frac{1}{3}$ in = 3 feet. It is 7 in on the map from the gift shop to the elevators. How many feet does that translate to?

   A 7 feet
   B 63 feet
   C 24 feet
   D 60 feet

8. If $x = 5$ and $y = 7$, what is y when $x = 20$ using direct variation?

   A 21
   B 28
   C 35
   D 40

9. Ashley's cat, Scratches, had kittens. Three of the kittens were all white, and two were black and white. What is the ratio of black and white kittens to all white kittens?

   A $\frac{1}{2}$
   B $\frac{1}{5}$
   C $\frac{2}{3}$
   D $\frac{2}{5}$

10. The scale on a map of an arena is 1 inch = 100 feet. The length of the football field in the arena is 3 inches in length on the map. How long is the football field used by the football players?

   A 300 feet
   B 100 feet
   C 150 feet
   D 350 feet

Chapter 5
Algebra Word Problems

This chapter covers the following Alabama objectives and standards in mathematics:

	Objective(s)
Standard VI	1
Standard VII	8

5.1   Rate

Example 1:   Laurie traveled 312 miles in 6 hours. What was her average rate of speed?

Divide the number of miles by the number of hours. $\dfrac{312 \text{ miles}}{6 \text{ hours}} = 52$ miles/hour

Laurie's average rate of speed was 52 miles per hour (or 52 mph).

Find the average rate of speed in each problem below.

1. A race car went 500 miles in 4 hours. What was its average rate of speed?

2. Carrie drove 124 miles in 2 hours. What was her average speed?

3. After 7 hours of driving, Chad had gone 364 miles. What was his average speed?

4. Anna drove 360 miles in 8 hours. What was her average speed?

5. After 3 hours of driving, Paul had gone 183 miles. What was his average speed?

6. Nicole ran 25 miles in 5 hours. What was her average speed?

7. A train traveled 492 miles in 6 hours. What was its average rate of speed?

8. A commercial jet traveled 1,572 miles in 3 hours. What was its average speed?

9. Jillian drove 195 miles in 3 hours. What was her average speed?

10. Greg drove 8 hours from his home to a city 336 miles away. At what average speed did he travel?

11. Caleb drove 128 miles in two hours. What was his average speed in miles per hour?

12. After 9 hours of driving, Kate had traveled 405 miles. What speed did she average?

5.2   More Rates

Rates are often discussed in terms of miles per hour, but a rate can be any measured quantity divided by another measurement such as feet per second, kilometers per minute, mass per unit volume, etc. A rate can be how fast something is done. For example, a bricklayer may lay 80 bricks per hour. Rates can also be used to find measurements such as density. For example, 35 grams of salt in 1 liter of water gives the mixture a density of 35 grams/liter.

Example 2:   Nathan entered his snail in a race. His snail went 18 feet in 6 minutes. How fast did his snail move?

In this problem, the units given are feet and minutes, so the rate will be feet per minute (or feet/minute).

You need to find out how far the snail went in one minute.

$$\text{Rate equals } \frac{\text{distance}}{\text{time}} \text{ so } \frac{18 \text{ feet}}{6 \text{ minutes}} = \frac{3 \text{ feet}}{1 \text{ minute}}$$

Nathan's snail went an average of 3 feet per minute or $3\dfrac{\text{ft}}{\text{min}}$.

Find the average rate for each of the following problems.

1. Tewanda read a 2, 000-word news article in 8 minutes. How fast did she read the news article?

2. Chandler rides his bike to school every day. He travels 2, 560 feet in 640 seconds. How many feet did he travel per second?

3. Mr. Molier is figuring out the semester averages for his history students. He can calculate the average for 20 students in an hour. How long does it take him to figure the average for each student?

4. In 1908, John Hurlinger of Austria walked 1, 400 kilometers from Vienna to Paris on his hands. The journey took 55 days. What was his average speed per day?

5. Spectators at the Super Circus were amazed to watch a cannon shoot a clown 212 feet into a net in 4 seconds. How many feet per second did the clown travel?

6. Marcus Page, star receiver for the Big Bulls, was awarded a 5-year contract for 105 million dollars. How much will his annual rate of pay be if he is paid the same amount each year?

7. Duke Delaney scored 28 points during the 4 quarters of the basketball playoffs. What was his average score per quarter?

8. The new burger restaurant in Moscow serves 11, 208 customers during a 24-hour period. What is the average number of customers served per hour?

5.3   Algebra Word Problems

An equation states that two mathematical expressions are equal. In working with word problems, the words that mean equal are **equals, is, was, is equal to, amounts to,** and other expressions with the same meaning. To translate a word problem into an algebraic equation, use a variable to represent the unknown or unknowns you are looking for.

In the following example, let n be the number you are looking for.

Example 3:   Four more than twice a number is two less than three times the number.

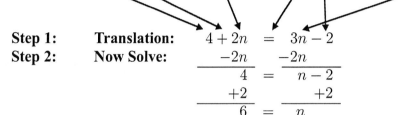

Step 1:   **Translation:** $4 + 2n = 3n - 2$
Step 2:   **Now Solve:**

$$\begin{array}{rcr} 4 + 2n & = & 3n - 2 \\ -2n & & -2n \\ \hline 4 & = & n - 2 \\ +2 & & +2 \\ \hline 6 & = & n \end{array}$$

The number is 6.
Substitute the number back into the original equation to check.

Translate the following word problems into equations and solve.

1. Four less than twice a number is ten. Find the number.

2. Three more than three times a number is one less than two times the number. What is the number?

3. The sum of seven times a number and the number is 24. What is the number?

4. Negative 18 is the sum of five and a number. Find the number.

5. Negative 14 is equal to ten minus the product of six and a number. What is the number?

6. Two less than twice a number equals the number plus 12. What is the number?

7. The difference between three times a number and 31 is two. What is the number?

8. Sixteen is fourteen less than the product of a number and five. What is the number?

9. Eight more than twice a number is four times the difference between five and the number. What is the number?

10. Three less than twice a number is three times the sum of one and the number. What is the number?

5.4   Real-World Linear Equations

Linear equations are very useful mathematical tools. They allow us to show relationships between two variables.

Example 4:   A local cell phone company uses the equation $y = \frac{5}{2}x + 10$ to determine the charges for usage where $y =$ the cost and $x =$ the minutes used. How much will Jessica's bill be if she talked for 40 minutes?

Step 1:   Substitute the known value in for x.
$y = \frac{5}{2}(40) + 10$

Step 2:   Simplify.
$y = 100 + 10 = 110$
Jessica's bill will be $110.

Example 5:   Vincent bought a luxury car for $165,000$ and its value has depreciated linearly. After 5 years the value was $137,000$. What is the amount of yearly depreciation?

Step 1:   First find how much the car's value depreciated in 5 years.
$165,000 - 137,000 = 28,000$

Step 2:   Next, find the yearly depreciation by dividing $28,000$ by the amount of years, 5.
$28,000 \div 5 = 5,600$
The value of Vincent's car depreciated $5,600$ each year.

Example 6:   In 1990, the average cost of a new house was $123,000$. By the year 2000, the average cost of a new house was $134,150$. Based on a linear model, what is the predicted average cost for 2008?

Step 1:   First, we need to find the difference between the average cost of a new house in the year 1990 and the average cost of a new house in the year 2000.
$134,150 - 123,000 = 11,150$

Step 2:   Next, we need to find how much the average cost of a new house went up each year. Since it had been 10 years, divide the difference between the value in 2000 and 1990 by 10.
$11,150 \div 10 = 1,115$

Step 3:   Multiply the amount the average cost of a new house went up each year by the number of years between 2000 and 2008.
$1,115 \times 8 = 8,920$

Step 4:   Lastly, add the average cost of a new house in the year 2000 with the amount found in step 3.
$134,150 + 8,920 = 143,070$
$143,070$ is the predicted average cost of a new house for 2008.

Solve the following problems.

1. Acacia bought an MP3 player at Everywhere Electronics for $350 and its valued depreciated linearly. Three years later, she saw the same MP3 player at Everywhere Electronics for $125. What is the amount of yearly depreciation of Acacia's MP3 player?

2. Dustin bought a boat 10 years ago for $10,000. Its value depreciated linearly and now it is worth $2,500. What is the amount of yearly depreciation of Dustin's boat?

3. A small plane costs $500,000 new. Twenty years later it is valued at $150,000. Assuming a linear depreciation, what was the value of the plane when it was 14 years old?

4. In 1980, the price of a scientific calculator was $155. In 2005, the price was $15 dollars. Assuming the change in price was linear, what was the price of a scientific calculator in 1997?

5. In 1997, Justin bought a house for $120,000. In 2004, his house was worth $176,000. Based on a linear model, how much was Justin's house worth in 2001?

6. The attendance on the first day of the Sunny Day Festival was 325 people. The attendance on the third day was 382 people. Assuming the attendance will increase linearly each day, how many people will attend the Sunny Day Festival on the seventh day?

7. Two years ago Juanita bought 2 shirts for $15 and last year she bought 4 shirts for $45. Assuming the price will increase linearly, how much will 8 shirts cost Juanita this year?

8. In 1985, the average price of a new car was $9,000. In 2000, the average price was $24,750. Based on a linear model, what is the predicted average price for 2009?

Use the following information for questions 9–10.

Abbey is looking for a new cell phone provider. In her search, she has found 3 local companies: Gift of Gab, On the Go, and Connect. To determine their monthly charges, the 3 companies use the following equations.

Gift of Gab: $y = \frac{3}{4}x + 20$

On the Go: $y = \frac{1}{2}x + 60$

Connect: $y = 6x - 100$

9. Which is the cheapest provider if Abbey uses 200 minutes per month?

10. What if she used 100 minutes?

5.5   Word Problems with Formulas

The perimeter of a geometric figure is the distance around the outside of the figure.

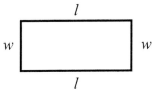

perimeter = $2l + 2w$

perimeter = $a + b + c$

Example 7:   The perimeter of a rectangle is 44 feet. The length of the rectangle is 6 feet more than the width. What is the measure of the width?

Step 1:   Let the variable be the length of the unknown side.
width = w length = $6 + w$

Step 2:   Use the equation for the perimeter of a rectangle as follows:
$2l + 2w$ = perimeter of a rectangle
$2(w + 6) + 2w = 44$

Step 3:   Solve for w.

Solution: width = 8 feet

Example 8:   The perimeter of a triangle is 26 feet. The second side is twice as long as the first. The third side is 1 foot longer than the second side. What is the length of the 3 sides?

Step 1:   Let x = first side $2x$ = second side $2x + 1$ = third side

Step 2:   Use the equation for perimeter of a triangle as follows:
sum of the length of the sides = perimeter of a triangle.
$x + 2x + 2x + 1 = 26$

Step 3:   Solve for x. $5x + 1 = 26$ so $x = 5$

Solution: first side $x = 5$ second side $2x = 10$ third side $2x + 1 = 11$

Solve the following word problems.

1. The length of a rectangle is 4 times longer than the width. The perimeter is 30. What is the width?

2. The length of a rectangle is 3 more than twice the width. The perimeter is 36. What is the length?

3. The perimeter of a triangle is 18 feet. The second side is two feet longer than the first. The third side is two feet longer then the second. What are the lengths of the sides?

4. In an isosceles triangle, two sides are equal. The third side is two less than twice the length of the sum of the two sides. The perimeter is 40. What are the lengths of the three sides?

5. The sum of the measures of the angles of a triangle is 180°. The second angle is three times the measure of the first angle. The third angle is four times the measure of the second angle. Find the measure of each angle.

6. The sum of the measures of the angles of a triangle is 180°. The second angle of a triangle is twice the measure of the first angle. The third angle is 20 more than 5 times the first. What are the measures of the three angles?

5.6   Age Problems

Example 9:   Tara is twice as old as Gwen. Their sister, Amy, is 5 years older than Gwen. If the sum of their ages is 29 years, find each of their ages.

Step 1:   We want to find each of their ages so there are three unknowns. Tara is twice as old as Gwen, and Amy is older than Gwen, so Gwen is the youngest. Let x be Gwen's age. From the problem we can see that:

$$\left. \begin{array}{rcl} \text{Gwen} & = & x \\ \text{Tara} & = & 2x \\ \text{Amy} & = & x+5 \end{array} \right\} \text{The sum of their ages is 29.}$$

Step 2:   Set up the equation, and solve for x.

$$\begin{array}{rcl} x + 2x + x + 5 & = & 29 \\ 4x + 5 & = & 29 \\ 4x & = & 29 - 5 \\ x & = & \dfrac{24}{4} \\ x & = & 6 \end{array}$$

Solution:
$$\begin{array}{rcl} \text{Gwen's age } (x) & = & 6 \\ \text{Tara's age } (2x) & = & 12 \\ \text{Amy's age } (x+5) & = & 11 \end{array}$$

Solve the following age problems.

1. Carol is 15 years older than her cousin Amanda. Cousin Bill is 4 times as old as Amanda. The sum of their ages is 99. Find each of their ages.

2. Derrick is 5 less than twice as old as Brandon. The sum of their ages is 31. How old are Derrick and Brandon?

3. Beth's mom is 5 times older than Beth. Beth's dad is 8 years older than Beth's mom. The sum of their ages is 74. How old are each of them?

4. Delores is 4 years more than three times as old as her son, Raul. If the difference between their ages is 34, how old are Delores and Raul?

5. Eileen is 9 years older than Karen. John is three times as old as Karen. The sum of their ages is 64. How old are Eileen, Karen, and John?

6. Taylor is 20 years younger than Jim. Andrew is twice as old as Taylor. The sum of their ages is 32. How old are Taylor, Jim, and Andrew?

The following problems work in the same way as the age problems. There are two or three items of different weight, distance, number, or size. You are given the total and asked to find the amount of each item.

7. Three boxes have a total height of 720 pounds. Box A weighs twice as much as Box B. Box C weighs 30 pounds more than Box A. How much do each of the boxes weigh?

8. There are 170 students registered for American History classes. There are twice as many students registered in second period as first period. There are 10 less than three times as many students registered in third period as in first period. How many students are in each period?

9. Mei earns $4 less than three times as much as Olivia. Shane earns twice as much as Mei. Together they earn $468 per week. How much does each person earn per week?

10. Ellie, the elephant, eats 4 times as much as Popcorn, the pony. Zac, the zebra, eats twice as much as Popcorn. Altogether, they eat 238 kilograms of feed per week. How much feed does each of them require each week?

11. The school cafeteria served three kinds of lunches today to 117 students. The students chose the cheeseburgers three times more often than the grilled cheese sandwiches. There were twice as many grilled cheese sandwiches sold as fish sandwiches. How many of each lunch were served?

12. Three friends drove southeast to New Mexico. Kyle drove half as far as Jamaal. Conner drove 4 times as far as Kyle. Altogether, they drove 476 miles. How far did each friend drive?

13. Bianca is taking collections for this year's Feed the Hungry Project. She has collected $300 more from Company A than from Company B and $700 more from Company C than from Company A. So far, she has collected $4, 300. How much did Company C give?

14. For his birthday, Torin got $50.00 more from his grandmother than from his uncle. His uncle gave him $10.00 less than his cousin. Torin received $135.00 in total. How much did he receive from his cousin?

15. Cassidy loves black and yellow jelly beans. She noticed when she was counting them that she had 7 less than three times as many black jelly beans as she had yellow jelly beans. In total, she counted 225 jelly beans. How many black jelly beans did she have?

16. Mrs. Vargus planted a garden with red and white rose bushes. Because she was studying to be a botanist, she counted the number of blossoms on each bush. She counted 4 times as many red blossoms as white blossoms. In total, she counted 1, 420 blossoms. How many red blossoms did she count?

5.7   Consecutive Integer Problems

	Examples:	Algebraic notation:
Consecutive integers follow each other in order	$1, 2, 3, 4$ $-3, -4, -5, -6$	$n, n+1, n+2, n+3$
Consecutive **even** integers:	$2, 4, 6, 8, 10$ $-12, -14, -16, -18$	$n, n+2, n+4, n+6$
Consecutive **odd** integers:	$3, 5, 7, 9$ $-5, -7, -9, -11$	$n, n+2, n+4, n+6$

Example 10:   The sum of three consecutive odd integers is 63. Find the integers.

Step 1:   Represent the three odd integers:
Let n = the first odd integer
$n + 2$ = the second odd integer
$n + 4$ = the third odd integer

Step 2:   The sum of the integers is 63, so the algebraic equation is
$n + n + 2 + n + 4 = 63$.   Solve for n.
$n = 19$

Solution: the first odd integer $= 19$
the second odd integer $= 21$
the third odd integer $= 23$

Check: Does $19 + 21 + 23 = 63$? Yes, it does.

Solve the following problems.

1. Find three consecutive even integers whose sum is 120.

2. Find three consecutive integers whose sum is -30.

3. The sum of three consecutive odd integers is 51. What are the numbers?

4. Find two consecutive odd integers such that five times the first equals three times the second.

5. Find two consecutive even integers such that seven times the first equals six times the second.

6. Find two consecutive odd numbers whose sum is eighty.

5.8 Inequality Word Problems

Inequality word problems involve staying under a limit or having a minimum goal one must meet.

Example 11: A contestant on a popular game show must earn a minimum of 800 points by answering a series of questions worth 40 points each per category in order to win the game. The contestant will answer questions from each of four categories. Her results for the first three categories are as follows: 160 points, 200 points, and 240 points. Write an inequality which describes how many points, (p), the contestant will need on the last category in order to win.

Step 1: Add to find out how many points she already has. $160 + 200 + 240 = 600$

Step 2: Subtract the points she already has from the minimum points she needs. $800 - 600 = 200$. She must get at least 200 points in the last category to win. If she gets more than 200 points, that is okay, too. To express the number of points she needs, use the following inequality statement:

$p \geq 200$ The points she needs must be greater than or equal to 200.

Solve each of the following problems using inequalities.

1. Stella wants to place her money in a high interest money market account. However, she needs at least $1,500 to open an account. Each month, she sets aside some of her earnings in a savings account. In January through June, she added the following amounts to her savings: $145, $203, $210, $120, $102, and $115. Write an inequality which describes the amount of money she can set aside in July to qualify for the money market account.

2. A high school band program will receive $2,000.00 for selling $12,000.00 worth of coupon books. Six band classes participate in the sales drive. Classes 1–5 collect the following amounts of money: $2,400, $2,800, $1,500, $2,320, and $2,550. Write an inequality which describes the amount of money the sixth class must collect so that the band will receive $2,000.

3. A small elevator has a maximum capacity of 1,200 pounds before the cable holding it in place snaps. Six people get on the elevator. Five of their weights follow: 120, 240, 150, 215, and 170. Write an inequality which describes the amount the sixth person can weigh without snapping the cable.

4. A small high school class of 9 students were told they would receive a pizza party if their class average was 90% or higher on the next exam. Students 1–8 scored the following on the exam: 84, 95, 99, 87, 92, 93, 100, and 98. Write an inequality which describes the score the ninth student must make for the class to qualify for the pizza party.

5. Raymond wants to spend his entire credit limit on his credit card. His credit limit is $3,000. He purchases items costing $750, $1,120, $42, $159, $8, and $71. Write an inequality which describes the amounts Raymond can put on his credit card for his next purchases.

5.9   Understanding Simple Interest

I = PRT is a formula to figure out the **cost of borrowing money** or the **amount you earn** when you **put money in a savings account**. For example, when you want to buy a used truck or car, you go to the bank and borrow the $7,000 you need. The bank will charge you **interest** on the $7,000. If the simple interest rate is 9% for four years, you can figure the cost of the interest with this formula.

First, you need to understand these terms:

I = Interest = The amount charged by the bank or other lender
P = Principal = The amount you borrow
R = Rate = The interest the bank is charging you
T = Time = How many years you will take to pay off the loan

Example 12:   In the problem above: **I = PRT**. This means the **interest** equals the **principal**, times the **rate**, times the **time** in **years**.

$$I = \$7,000 \times 9\% \times 4 \text{ years}$$
$$I = \$7,000 \times 0.09 \times 4$$
$$I = \$2,520$$

Use the formula I = PRT to work the following problems.

1. Craig borrowed $1,800 from his parents to buy a stereo. His parents charged him 3% simple interest for 2 years. How much interest did he pay his parents?

2. Raul invested $5,000 in a savings account that earned 2% simple interest. If he kept the money in the account for 5 years, how much interest did he earn?

3. Bridgette borrowed $11,000 to buy a car. The bank charged 12% simple interest for 7 years. How much interest did she pay the bank?

4. A tax accountant invested $25,000 in a money market account for 3 years. The account earned 5% simple interest. How much interest did the accountant make on his investment?

5. Linda Kay started a savings account for her nephew with $2,000. The account earned 6% simple interest. How much interest did the account accumulate in 6 months?

6. Renada bought a living room set on credit. The set sold for $2,300, and the store charged her 9% simple interest for three months. How much interest did she pay?

7. Duane took out a $3,500 loan at 8% simple interest for 3 years. How much interest did he pay for borrowing the $3,500?

5.10   Tips and Commissions

Vocabulary

Tip: A **tip** is money given to someone doing a service for you such as a server,
                  hair stylist, porter, cab driver, grocery bagger, etc.

Commission: In many businesses, sales people are paid on **commission** — a percent of the total
                  sales they make.

Problems requiring you to figure a tip, commission, or percent of a total are all done in the same
way.

Example 13:   Ramon makes a 4% commission on an $8,000 pickup truck he sold. How much
                  is his commission?

$$\begin{array}{r} \textbf{TOTAL COST} \\ \times\ \underline{\textbf{RATE OF COMMISSION}} \\ \textbf{COMMISSION} \end{array} \qquad \begin{array}{r} \$8,000 \\ \times\ \underline{\ 0.04\ } \\ \$320.00 \end{array}$$

Solve each of the following problems.

1. Mia makes a 12% commission on all her sales. This month she sells $9,000 worth of
   merchandise. What is her commission?

2. Marcus gives 25% of his income to his parents to help cover expenses. He earns $340 per
   week. How much money does he give his parents?

3. Jan pays $640 per month for rent. If rent goes up by 5%, how much can Jan expect to pay
   monthly next year?

4. The total bill at Jake's Catfish Place comes to $35.80. Palo wants to leave a 15% tip. How
   much money will he have to leave for the tip?

5. Rami makes $2,400 per month and puts 6% in a savings account. How much does he save per
   month?

6. Christina makes $2,550 per month. Her boss promises her a 7% raise. How much more will
   she make per month?

7. Out of 150 math students, 86% pass. How many students pass math class?

8. Marta sells Sue Anne Cosmetics and gets a 20% commission on all her sales. Last month, she
   sold $560.00 worth of cosmetics. How much was her commission?

Chapter 5 Review

Solve each of the following problems.

1. Beth is four more than three times older than Ted. The sum of their ages is 60. How old is Ted?

2. The band members sold tickets to their concert performance. Some were $3 tickets, and some were $8 tickets. There were 10 more than twice as many $8 tickets sold as $3 tickets. The total sales were $1,448. How many tickets of each price were sold?

3. Three consecutive integers have a sum of 54. Find the integers.

4. One number is 8 more than the other number. Twice the smaller number is 7 more than the larger number. What are the numbers?

5. The perimeter of a triangle is 48 inches. The second side is four inches longer than the first side. The third side is one inch longer than the second. Find the length of each side.

6. Joe, Craig, and Dylan have a combined weight of 326 pounds. Craig weighs 40 pounds more than Joe. Dylan weighs 12 pounds more than Craig. How many pounds does Craig weigh?

7. Lena and Jodie are sisters, and together they have 56 bottles of nail polish. Lena bought 4 more than half the bottles. How many did Jodie buy?

8. With the $5,000 he has to spend, Jim buys 5,000 carnations at $0.30 each, 4,000 tulips at $0.60 each, and 300 irises at $0.25 each. Write an inequality which describes how many roses, r, Jim can buy if roses cost $0.80 each.

9. Mr. Chan purchased the 90 shares for $0.50 each last month, and the shares are now worth $3.80 each. Write an inequality which describes how much profit, p, Mr. Chan can make by selling his shares.

10. The Jones family traveled 300 miles in 5 hours. What was their average speed?

11. Last year Rikki sang 960 songs with his rock band. How many songs did he sing per month?

12. Erin is looking for a new job. During her interviews, Company A says pay is determined by the equation $y = 13x - 12$, where x is the number of hours worked. How much will Erin make if she can only work ten hours at this company?

13. Cameron lives in Woodstock, Georgia. In his research, he found that the population was $10,050$ in the year 2000 and in 2007, the population was $23,000$. What is the predicted population for 2014?

14. Timothy bought a car for $5,500$ in 1975. In 2000, he learned that, fully restored, his car was worth $65,000$. Based on a linear model, how much was Timothy's car worth in 2006? How much will his car be worth in 2009?

15. Celeste makes 6% commission on her sales. If her sales for a week total $4,580, what is her commission?

16. McMartin's is offering a deal on fitness club memberships. You can pay $999 up front for a 3-year membership, or pay $200 down and $30 per month for 36 months. How much would you save by paying up front?

17. Jeneane earned $340.20 commission by selling $5670 worth of products. What percent commission did she earn?

18. Tara put $500 in a savings account that earned 3% simple interest. How much interest did she make after 5 years?

19. Linda took out a simple interest loan for $7,000 at 11% interest for 5 years. How much interest did she have to pay back?

20. Tyler brought in $25,000 worth of sales in the last 10 days. He earns 15% commission on his sales. How much was his commission?

Chapter 5 Test

1. Ross is five years older than twice his sister Holly's age. The difference is their ages is 14 years. How old is Holly?

   A 9
   B 23
   C 3
   D 18

2. The sum of two numbers is 27. The larger number is 6 more than twice the smaller number. What are the numbers?

   A 11, 16
   B 19, 8
   C 7, 20
   D 3, 24

3. The perimeter of a rectangle is 292 feet. The length of the rectangle is 4 feet less than 5 times the width. What is the length and width of the rectangle?

   A length = 121, width = 25
   B length = 114.3, width = 23.7
   C length = 25, width = 121
   D length = 121.7, width = 24.3

4. Janet and Artie want to play tug-of-war. Artie pulls with 200 pounds of force while Janet pulls with 60 pounds of force. In order to make this a fair contest, Janet enlists the help of her friends Trudi, Sherri, and Bridget who pull with 20, 25, and 55 pounds respectively. Write an inequality describing the minimum amount Janet's fourth friend, Tommy, must pull to beat Artie.

   A $x > 40$ pounds of force
   B $x < 40$ pounds of force
   C $x > 100$ pounds of force
   D $x < 100$ pounds of force

5. Jesse and Larry entered a pie eating contest. Jesse ate 5 less than twice as many pies as Larry. They ate a total of 16 pies. How many pies did Larry eat?

   A 3.7 pies
   B 9 pies
   C 21 pies
   D 7 pies

6. There is a new bike that Bianca has had her eye on for a few weeks. The bike costs $75. Her allowance is 10 dollars per week. If she saves 60% of her allowance each week, write an inequality that describes the minimum amount of weeks, y, that Bianca must save in order to buy that bike.

   A $y > 75 - 0.6\,(10)$
   B $y > 45\,(10)$
   C $y > \dfrac{75}{0.6\,(10)}$
   D $10 > \dfrac{75}{0.6y}$

7. The sum of two numbers is fourteen. The sum of six times the smaller number and two equals four less than the product of three and the larger number. Find the two numbers.

   A 6 and 8
   B 5 and 9
   C 3 and 11
   D 4 and 10

8. Find three consecutive odd numbers whose sum is three hundred three.

   A 100, 101, 102
   B 99, 101, 103
   C 99, 103, 107
   D 99, 100, 101

9. Tracie and Marcia drove to northern California to see Marcia's sister in Eureka. Tracie drove one hour more than four times as much as Marcia. The trip took a total of 21 driving hours. How many hours did Tracie drive?

A 17 hours
B 4 hours
C 20 hours
D 5 hours

10. Alisha climbed a mountain that was 4,760 feet high in 14 hours. What was her average speed per hour?

A 476 ft/hr
B 4,774 ft/hr
C 340 ft/hr
D 66,640 ft/hr

11. The new school copy machine makes 3,480 copies per hour. How many copies does this machine make per minute?

A 58 copies/minute
B 56.8 copies/minute
C 208,800 copies/minute
D 5 copies/minute

12. Connie drove for 2 hours at a constant speed of 55 mph. How many total miles did she travel?

A 27.5 miles
B 110 miles
C 220 miles
D 490 miles

13. A hat you want costs $17.00 and you buy it on sale for 30% off, how much will you save?

A $3.40
B $51.00
C $34.00
D $5.10

14. Using the formula **I = PRT**, how much interest will you pay on a loan for $5,000 at 7% interest for 2 years?

A $350
B $700
C $140
D $190

15. Using the formula **I = PRT**, how much interest will you earn on $300 at 3% interest for 3 years?

A $2.70
B $90
C $9
D $27

16. Will got a 25% tip on a meal costing $14.00. What was the amount of the tip?

A $3.50
B $3.00
C $2.50
D $2.80

Chapter 6
Polynomials

This chapter covers the following Alabama objectives and standards in mathematics:

	Objective(s)
Standard I	2, 3

Polynomials are algebraic expressions which include **monomials** containing one term, **binomials** which contain two terms, and **trinomials**, which contain three terms. Expressions with more than three terms are called **polynomials. Terms** are separated by plus and minus signs.

Examples

Monomials	Binomials	Trinomials	Polynomials
$4f$	$4t + 9$	$x^2 + 2x + 3$	$x^3 - 3x^2 + 3x - 9$
$3x^3$	$9 - 7g$	$5x^2 - 6x - 1$	$p^4 + 2p^3 + p^2 - 5 + p9$
$4g^2$	$5x^2 + 7x$	$y^4 + 15y^2 + 100$	
2	$6x^3 - 8x$		

6.1 Adding and Subtracting Monomials

Two **monomials** are added or subtracted as long as the **variable and its exponent** are the **same**. This is called combining like terms. Use the same rules you used for adding and subtracting integers.

Example 1:
$$4x + 5x = 9x$$

$$\begin{array}{r} 3x^4 \\ -8x^4 \\ \hline -5x^4 \end{array}$$

$$2x^2 - 9x^2 = -7x^2$$

$$\begin{array}{r} 5y \\ +2y \\ \hline 7y \end{array}$$

$$6y^3 - 5y^3 = y^3$$

Remember: When the integer in front of the variable is "1", it is usually not written. $1x^2$ is the same as x^2, and $-1x$ is the same as $-x$.

Add or subtract the following monomials.

1. $2x^2 + 5x^2$

2. $5t + 8t$

3. $9y^3 - 2y^3$

4. $6g - 8g$

5. $7y^2 + 8y^2$

6. $s^5 + s^5$

7. $-2x - 4x$

8. $4w^2 - w^2$

9. $z^4 + 9z^4$

10. $-k + 2k$

11. $3x^2 - 5x^2$

12. $9t + 2t$

13. $-7v^3 + 10v^3$

14. $-2x^3 + x^3$

15. $10y^4 - 5y^4$

16.	y^4 $+2y^4$	18.	$8t^2$ $+7t^2$	20.	$5w^2$ $+8w^2$	22.	$-5z$ $+9z$	24.	$7t^3$ $-6t^3$

17.	$4x^3$ $-9x^3$	19.	$-2y$ $-4y$	21.	$11t^3$ $-4t^3$	23.	$4w^5$ $+w^5$	25.	$3x$ $+8x$

6.2 Adding Polynomials

When adding **polynomials,** make sure the exponents and variables are the same on the terms you are combining. The easiest way is to put the terms in columns with **like exponents** under each other. Each column is added as a separate problem. Fill in the blank spots with zeros if it helps you keep the columns straight. You never carry to the next column when adding polynomials.

Example 2: Add $3x^2 + 14$ and $5x^2 + 2x$

$$\begin{array}{r} 3x^2 + 0x + 14 \\ (+)\,5x^2 + 2x + 0 \\ \hline 8x^2 + 2x + 14 \end{array}$$

Example 3: $(4x^3 - 2x) + (-x^3 - 4)$

$$\begin{array}{r} 4x^3 - 2x + 0 \\ (+)\,-x^3 + 0x - 4 \\ \hline 3x^3 - 2x - 4 \end{array}$$

Add the following polynomials.

1. $y^2 + 3y + 2$ and $2y^2 + 4$

2. $(5y^2 + 4y - 6) + (2y^2 - 5y + 8)$

3. $5x^3 - 2x^2 + 4x - 1$ and $3x^2 - x + 2$

4. $-p + 4$ and $5p^2 - 2p + 2$

5. $(w - 2) + (w^2 + 2)$

6. $4t^2 - 5t - 7$ and $8t + 2$

7. $t^4 + t + 8$ and $2t^3 + 4t - 4$

8. $(3s^3 + s^2 - 2) + (-2s^3 + 4)$

9. $(-v^2 + 7v - 8) + (4v^3 - 6v + 4)$

10. $6m^2 - 2m + 10$ and $m^2 - m - 8$

11. $-x + 4$ and $3x^2 + x - 2$

12. $(8t^2 + 3t) + (-7t^2 - t + 4)$

13. $(3p^4 + 2p^2 - 1) + (-5p^2 - p + 8)$

14. $12s^3 + 9s^2 + 2s$ and $s^3 + s^2 + s$

15. $(-9b^2 + 7b + 2) + (-b^2 + 6b + 9)$

16. $15c^2 - 11c + 5$ and $-7c^2 + 3c - 9$

17. $5c^3 + 2c^2 + 3$ and $2c^3 + 4c^2 + 1$

18. $-14x^3 + 3x^2 + 15$ and $7x^3 - 12$

19. $(-x^2 + 2x - 4) + (3x^2 - 3)$

20. $(y^2 - 11y + 10) + (-13y^2 + 5y - 4)$

21. $3d^5 - 4d^3 + 7$ and $2d^4 - 2d^3 - 2$

22. $(6t^5 - t^3 + 17) + (4t^5 + 7t^3)$

23. $4p^2 - 8p + 9$ and $-p^2 - 3p - 5$

24. $20b^3 + 15b$ and $-4b^2 - 5b + 14$

25. $(-2w + 11) + (w^3 + w - 4)$

26. $(25z^2 + 13z + 8) + (z^2 - 2z - 10)$

6.3   Subtracting Polynomials

When you subtract polynomials, it is important to remember to change all the signs in the subtracted polynomial (the subtrahend) and then add.

Example 4: $(4y^2 + 8y + 9) - (2y^2 + 6y - 4)$

    Step 1:   Copy the subtraction problem into vertical form. Make sure you line up the terms with like exponents under each other.

$$\begin{array}{r} 4y^2 + 8y + 9 \\ (-)\,2y^2 + 6y - 4 \\ \hline \end{array}$$

    Step 2:   Change the subtraction sign to addition and all the signs of the subtracted polynomial to the opposite sign.

$$\begin{array}{r} 4y^2 + 8y + 9 \\ (+)\,-2y^2 - 6y + 4 \\ \hline 2y^2 + 2y + 13 \end{array}$$

Subtract the following polynomials.

1. $(2x^2 + 5x + 2) - (x^2 + 3x + 1)$

2. $(8y - 4) - (4y + 3)$

3. $(11t^3 - 4t^2 + 3) - (-t^3 + 4t^2 - 5)$

4. $(-3w^2 + 9w - 5) - (-5w^2 - 5)$

5. $(6a^5 - a^3 + a) - (7a^5 + a^2 - 3a)$

6. $(14c^4 + 20c^2 + 10) - (7c^4 + 5c^2 + 12)$

7. $(5x^2 - 9x) - (-7x^2 + 4x + 8)$

8. $(12y^3 - 8y^2 - 10) - (3y^3 + y + 9)$

9. $(-3h^2 - 7h + 7) - (5h^2 + 4h + 10)$

10. $(10k^3 - 8) - (-4k^3 + k^2 + 5)$

11. $(x^2 - 5x + 9) - (6x^2 - 5x + 7)$

12. $(12p^2 + 4p) - (9p - 2)$

13. $(-2m - 8) - (6m + 2)$

14. $(13y^3 + 2y^2 - 8y) - (2y^3 + 4y^2 - 7y)$

15. $(7g + 3) - (g^2 + 4g - 5)$

16. $(-8w^3 + 4w) - (-10w^3 - 4w^2 - w)$

17. $(12x^3 + x^2 - 10) - (3x^3 + 2x^2 + 1)$

18. $(2a^2 + 2a + 2) - (-a^2 + 3a + 3)$

19. $(c + 19) - (3c^2 - 7c + 2)$

20. $(-6v^2 + 12v) - (3v^2 + 2v + 6)$

21. $(4b^3 + 3b^2 + 5) - (7b^3 - 8)$

22. $(15x^3 + 5x^2 - 4) - (4x^3 - 4x^2)$

23. $(8y^2 - 2) - (11y^2 - 2y - 3)$

24. $(-z^2 - 5z - 8) - (3z^2 - 5z + 5)$

6.4   Multiplying Monomials

When two monomials have the **same variable**, you can multiply them. Then, add the **exponents** together. If the variable has no exponent, it is understood that the exponent is 1.

Example 5: $4x^4 \times 3x^2 = 12x^6$ $\qquad\qquad$ $2y \times 5y^2 = 10y^3$

Multiply the following monomials.

1. $6a \times 9a^5$
2. $2x^6 \times 5x^3$
3. $4y^3 \times 3y^2$
4. $10t^2 \times 2t^2$
5. $2p^5 \times 4p^2$
6. $9b^2 \times 8b$
7. $3c^3 \times 3c^3$

8. $2d^8 \times 9d^2$
9. $6k^3 \times 5k^2$
10. $7m^5 \times m$
11. $11z \times 2z^7$
12. $3w^4 \times 6w^5$
13. $4x^4 \times 5x^3$
14. $5n^2 \times 3n^3$

15. $8w^7 \times w$
16. $10s^6 \times 5s^3$
17. $4d^5 \times 4d^5$
18. $5y^2 \times 8y^6$
19. $7t^{10} \times 3t^5$
20. $6p^8 \times 2p^3$
21. $x^3 \times 2x^3$

When problems include negative signs, follow the rules for multiplying integers.

22. $-7s^4 \times 5s^3$
23. $-6a \times -9a^5$
24. $4x \times -x$
25. $-3y^2 \times -y^3$
26. $-5b^2 \times 3b^5$
27. $9c^4 \times -2c$
28. $-4t^3 \times 8t^3$

29. $10d \times -8d^7$
30. $-3g^6 \times -2g^3$
31. $-7s^4 \times 7s^3$
32. $-d^3 \times -2d$
33. $11p \times -2p^5$
34. $-5x^7 \times -3x^3$
35. $8z^4 \times 7z^4$

36. $-4w \times -5w^8$
37. $-5y^4 \times 6y^2$
38. $9x^3 \times -7x^5$
39. $-a^4 \times -a$
40. $-7k^2 \times 3k$
41. $-15t^2 \times -t^4$
42. $3x^8 \times 9x^2$

6.5   Multiplying Monomials by Polynomials

In the chapter on solving multi-step equations, you learned to remove parentheses by multiplying the number outside the parentheses by each term inside the parentheses: $2(4x - 7) = 8x - 14$. Multiplying monomials by polynomials works the same way.

Example 6: $-5t(2t^2 - 7t + 9)$

Step 1:   Multiply $-5t \times 2t^2 = -10t^3$

Step 2:   Multiply $-5t \times -7t = 35t^2$

Step 3:   Multiply $-5t \times 9 = -45t$

Step 4:   Arrange the answers horizontally in order: $-10t^3 + 35t^2 - 45t$

Remove parentheses in the following problems.

1. $3x(3x^2 + 4x - 1)$

2. $4y(y^3 - 7)$

3. $7a^2(2a^2 + 3a + 2)$

4. $-5d^3(d^2 - 5d)$

5. $2w(-4w^2 + 3w - 8)$

6. $8p(p^3 - 6p + 5)$

7. $-9b^2(-2b + 5)$

8. $2t(t^2 - 4t - 10)$

9. $10c(4c^2 + 3c - 7)$

10. $6z(2z^4 - 5z^2 - 4)$

11. $-9t^2(3t^2 + 5t + 6)$

12. $c(-3c - 5)$

13. $3p(p^3 - p^2 - 9)$

14. $-k^2(2k + 4)$

15. $-3(4m^2 - 5m + 8)$

16. $6x(-7x^3 + 10)$

17. $-w(w^2 - 4w + 7)$

18. $2y(5y^2 - y)$

19. $3d(d^5 - 7d^3 + 4)$

20. $-5t(-4t^2 - 8t + 1)$

21. $7(2w^2 - 9w + 4)$

22. $3y^2(y^2 - 11)$

23. $v^2(v^2 + 3v + 3)$

24. $8x(2x^3 + 3x + 1)$

25. $-5d(4d^2 + 7d - 2)$

26. $-k^2(-3k + 6)$

27. $3x(-x^2 - 5x + 5)$

28. $4z(4z^4 - z - 7)$

29. $-5y(9y^3 - 3)$

30. $2b^2(7b^2 + 4b + 4)$

6.6   Removing Parentheses and Simplifying

In the following problem, you must multiply each set of parentheses by the numbers and variables outside the parentheses, and then add the polynomials to simplify the expressions.

Example 7: $8x\left(2x^2 - 5x + 7\right) - 3x\left(4x^2 + 3x - 8\right)$

Step 1:   Multiply to remove the first set of parentheses.

$$8x\left(2x^2 - 5x + 7\right) = 16x^3 - 40x^2 + 56x$$

Step 2:   Multiply to remove the second set of parentheses.

$$-3x\left(4x^2 + 3x - 8\right) = -12x^3 - 9x^2 + 24x$$

Step 3:   Copy each polynomial in columns, making sure the terms with the same variable and exponent are under each other. Add to simplify.

$$\begin{array}{r} 16x^3 - 40x^2 + 56x \\ (+) -12x^3 - 9x^2 + 24x \\ \hline 4x^3 - 49x^2 + 80x \end{array}$$

Remove the parentheses and simplify the following problems.

1. $4t\left(t + 7\right) + 5t\left(2t^2 - 4t + 1\right)$

2. $-5y\left(3y^2 - 5y + 3\right) - 6y\left(y^2 - 4y - 4\right)$

3. $-3\left(3x^2 + 4x\right) + 5x\left(x^2 + 3x + 2\right)$

4. $2b\left(5b^2 - 8b - 1\right) - 3b\left(4b + 3\right)$

5. $8d^2\left(3d + 4\right) - 7d\left(3d^2 + 4d + 5\right)$

6. $5a\left(3a^2 + 3a + 1\right) - \left(-2a^2 + 5a - 4\right)$

7. $3m\left(m + 7\right) + 8\left(4m^2 + m + 4\right)$

8. $4c^2\left(-6c^2 - 3c + 2\right) - 7c\left(5c^3 + 2c\right)$

9. $-8w\left(-w + 1\right) - 4w\left(3w - 5\right)$

10. $6p\left(2p^2 - 4p - 6\right) + 3p\left(p^2 + 6p + 9\right)$

6.7   Multiplying Two Binomials Using the FOIL Method

When you multiply two binomials such as $(x + 6)(x - 5)$, you must multiply each term in the first binomial by each term in the second binomial. The easiest way is to use the **FOIL** method. If you can remember the word **FOIL**, it can help you keep order when you multiply. The "**F**" stands for **first**, "**O**" stands for **outside**, "**I**" stands for **inside**, and "**L**" stands for **last**.

F	**O**	**I**	**L**
FIRST	**OUTSIDE**	**INSIDE**	**LAST**
Multiply the **first** terms in each binomial	Next, multiply the **outside** terms.	Then, multiply the **inside** terms.	Last, multiply the **last** terms.

$$(x+6)(x-5) \qquad (x+6)(x-5) \qquad (x+6)(x-5) \qquad (x+6)(x-5)$$
$$x \times x = x^2 \qquad x \times -5 = -5x \qquad 6 \times x = 6x \qquad 6 \times -5 = -30$$
$$x^2 \quad + \quad -5x \quad + \quad 6x \quad + \quad -30$$

Now just combine like terms, $6x - 5x = x$, and write your answer.

$(x + 6)(x - 5) = x^2 + x - 30$.

Note: It is customary for mathematicians to write polynomials in descending order. That means that the term with the highest exponent comes first in a polynomial. The next highest exponent is second, and so on. When you use the **FOIL** method, the terms will always be in the customary order. You just need to combine like terms and write your answer.

1. $(y - 7)(y + 3)$
2. $(2x + 4)(x + 9)$
3. $(4b - 3)(3b - 4)$
4. $(6g + 2)(g - 9)$
5. $(7k - 5)(-4k - 3)$
6. $(8v - 2)(3v + 4)$
7. $(10p + 2)(4p + 3)$
8. $(3h - 9)(-2h - 5)$
9. $(w - 4)(w - 7)$
10. $(6x + 1)(x - 2)$
11. $(5t + 3)(2t - 1)$
12. $(4y - 9)(4y + 9)$
13. $(a + 6)(3a + 5)$
14. $(3z - 8)(z - 4)$
15. $(5c + 2)(6c + 5)$

16. $(y + 3)(y - 3)$
17. $(2w - 5)(4w + 6)$
18. $(7x + 1)(x - 4)$
19. $(6t - 9)(4t - 4)$
20. $(5b + 6)(6b + 2)$
21. $(2z + 1)(10z + 4)$
22. $(11w - 8)(w + 3)$
23. $(5d - 9)(9d + 9)$
24. $(9g + 2)(g - 2)$
25. $(4p + 7)(2p + 3)$
26. $(m + 5)(m - 5)$
27. $(8b - 8)(2b - 1)$
28. $(z + 3)(3z + 5)$
29. $(7y - 5)(y - 3)$
30. $(9x + 5)(3x - 1)$

31. $(3t + 1)(t + 10)$
32. $(2w - 9)(8w + 7)$
33. $(8s - 2)(s + 4)$
34. $(4k - 1)(8k + 9)$
35. $(h + 12)(h - 2)$
36. $(3x + 7)(7x + 3)$
37. $(2v - 6)(2v + 6)$
38. $(2x + 8)(2x - 3)$
39. $(k - 1)(6k + 12)$
40. $(3w + 11)(2w + 2)$
41. $(8y - 10)(5y - 3)$
42. $(6d + 13)(d - 1)$
43. $(7h + 3)(2h + 4)$
44. $(5n + 9)(5n - 5)$
45. $(6z + 5)(z - 8)$

6.8   Simplifying Expressions with Exponents

Example 8:   **Simplify** $(2a + 5)^2$
When you simplify an expression such as $(2a + 5)^2$, write
the expression as two binomials and use FOIL to simplify.
$(2a + 5)^2 = (2a + 5)(2a + 5)$
Using FOIL we have $4a^2 + 10a + 10a + 25 = 4a^2 + 20a + 25$

Example 9:   **Simplify** $4(3a + 2)^2$
Using order of operations, we must simplify the exponent first.

$4(3a + 2)^2$

$4(3a + 2)(3a + 2)$

$4(9a^2 + 6a + 6a + 4)$

$4(9a^2 + 12a + 4)$ Now multiply by 4.

$4(9a^2 + 12a + 4) = 36a^2 + 48a + 16$

Multiply the following binomials.

1. $(y + 3)^2$

2. $2(2x + 4)^2$

3. $6(4b - 3)^2$

4. $5(6g + 2)^2$

5. $(-4k - 3)^2$

6. $3(-2h - 5)^2$

7. $-2(8v - 2)^2$

8. $(10p + 2)^2$

9. $6(-2h - 5)^2$

10. $6(w - 7)^2$

11. $2(6x + 1)^2$

12. $(9x + 2)^2$

13. $(5t + 3)^2$

14. $3(4y - 9)^2$

15. $8(a + 6)^2$

16. $4(3z - 8)^2$

17. $3(5c + 2)^2$

18. $4(3t + 9)^2$

6.9   Absolute Value

The absolute value of a number is the distance the number is from zero on the number line.

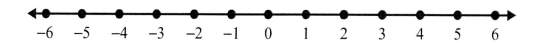

The absolute value if 6 is written $|6|$. $|6| = 6$

The absolute value of -6 is written $|-6|$. $|-6| = 6$

Both 6 and -6 are the same distance, 6 spaces, from zero so their absolute value is the same: 6.

Examples:

$|-4| = 4$ $\qquad\qquad$ $-|-4| = -4$ $\qquad\qquad$ $|-9| + 5 = 9 + 5 = 14$

$|9| - |8| = 9 - 8 = 1$ $\qquad$ $|6| - |-6| = 6 - 6 = 0$ $\qquad$ $|-5| + |-2| = 5 + 2 = 7$

Simplify the following absolute value problems.

1. $|9|$

2. $-|5|$

3. $|-25|$

4. $-|-12|$

5. $-|64|$

6. $|-2|$

7. $-|-3|$

8. $|-4| - |3|$

9. $|-8| - |-4|$

10. $|5| + |-4|$

11. $|-2| + |6|$

12. $|10| + |8|$

13. $|-2| + |4|$

14. $|-3| + |-4|$

15. $|7| - |-5|$

16. $|8| + |-12|$

6.10   Simplifying Absolute Value Expressions

When simplifying absolute value expressions, one must follow the rules of absolute values.

Example 10:   Multiply: $|6| \times |-8|$

Step 1:   Apply the absolute value rule, which states that the absolute value of any number, positive or negative, is that number as a positive.

$|6| = 6$ and $|-8| = 8$

Step 2:   Multiply. $6 \times 8 = 48$

Example 11:   Simplify: $3.2 - |-6.1 + 2.6|$

Step 1:   Apply the absolute value rules.

$|-6.1 + 2.6| = |-3.5| = 3.5$

Step 2:   Subtract. $3.2 - 3.5 = -0.3$

Solve the problems below using the rules for absolute values.

1. $|-4| + |2 - 5|$

2. $|12| - |-3|$

3. $\dfrac{|-14|}{|-2|}$

4. $-|15| \div |3|$

5. $|-5| - |7 + 2.4|$

6. $|-1 \times 5|$

7. $|1 + 7.4| \times |-4|$

8. $|3| (|2.1 + 9.6|)$

9. $|2.2 - 0.6| (-|5|)$

10. $3 + |-7|$

11. $|-12| \cdot |2 - 8|$

12. $\dfrac{-|-18|}{|-6|}$

13. $-21 \div |-7|$

14. $-|5 + 6| \times |3|$

15. $|-6| (|7|)$

16. $\dfrac{|-3 + 6|}{|-3|}$

17. $|-5.3 + 2.4| + |-8|$

18. $\dfrac{-|34|}{|-6 + 11|}$

19. $8 (-|4 - 5|)$

20. $|-1| + |-8|$

21. $|-7 + 11.1| - |-4|$

22. $|-2| + |2|$

23. $\dfrac{|18|}{|-6|}$

24. $|5.7 + 14.9| - |3|$

Chapter 6 Review

Simplify.

1. $3a^4 + 20a^4$

2. $(8x^4y^5)(20xy^7)$

3. $-6z^4(z + 3)$

4. $(5b^4)(7b^3)$

5. $8x^4 - 20x^4$

6. $(7p - 5) - (3p + 4)$

7. $-7t(3t + 20)^2$

8. $(3w^3y^4)(5wy^7)$

9. $3(4g + 3)^2$

10. $25d^5 - 20d^5$

11. $(8w - 5)(w - 9)$

12. $27t^4 + 5t^4$

13. $(8c^5)(20c^4)$

14. $(20x + 4)(x + 7)$

15. $5y(5y^4 - 20y + 4)$

16. $(9a^5b)(4ab^3)(ab)$

17. $(7w^6)(20w^{20})$

18. $9x^3 + 24x^3$

19. $27p^7 - 22p^7$

20. $(3s^5t^4)(5st^3)$

21. $(5d + 20)(4d + 8)$

22. $5w(-3w^4 + 8w - 7)$

23. $45z^6 - 20z^6$

24. $-8y^3 - 9y^3$

25. $(8x^5)(8x^7)$

26. $28p^4 + 20p^4$

27. $(a^4v)(4av)(a^3v^6)$

28. $5(6y - 7)^2$

29. $(3c^4)(6c^9)$

30. $(5x^7y^3)(4xy^3)$

31. Add $4x^4 + 20x$ and $7x^4 - 9x + 4$

32. $5t(6t^4 + 5t - 6) + 9t(3t + 3)$

33. Subtract $y^4 + 5y - 6$ from $3y^4 + 8$

34. $4x(5x^4 + 6x - 3) + 5x(x + 3)$

35. $(6t - 5) - (6t^4 + t - 4)$

36. $(5x + 6) + (8x^4 - 4x + 3)$

37. Subtract $7a - 4$ from $a + 20$

38. $(-4y + 5) + (5y - 6)$

39. $4t(t + 6) - 7t(4t + 8)$

40. Add $3c - 5$ and $c^4 - 3c - 4$

41. $4b(b - 5) - (b^4 + 4b + 2)$

42. $(6k^4 + 7k) + (k^4 + k + 20)$

43. $(q^4r^3)(3qr^4)(4q^5r)$

44. $(7df)(d^5f^4)(4df)$

45. $(8g^4h^3)(g^3h^6)(6gh^3)$

46. $(9v^4x^3)(3v^6x^4)(4v^5x^5)$

47. $(3n^4m^4)(20n^4m)(n^3m^8)$

48. $(22t^4a^4)(5t^3a^9)(4t^6a)$

49. $2b(b - 4) - (b^2 + 2b + 1)$

50. $|4|$

51. $|-6|$

52. $|-3.4 - 5.2| + |7.9|$

53. $|-2| \times |-6|$

54. $|8| - |5 + 2|$

Chapter 6 Test

1. $2x^2 + 5x^2 =$

   A $10x^4$
   B $7x^4$
   C $7x^2$
   D $10x^2$

2. $-8m^3 + m^3 =$

   A $-8m^6$
   B $-8m^9$
   C $-9m^6$
   D $-7m^3$

3. $(6x^3 + x^2 - 5) + (-3x^3 - 2x^2 + 4) =$

   A $3x^3 - x^2 - 1$
   B $3x^3 - 3x^2 - 1$
   C $3x^3 - 3x^2 - 9$
   D $-3x^3 - 3x^2 - 1$

4. $(-7c^2 + 5c + 3) + (-c^2 - 7c + 2) =$

   A $-3x^3 - 3x^2 - 1$
   B $-8c^2 - 2c + 5$
   C $-6c^2 - 12c + 5$
   D $-8c^2 - 12c + 5$

5. $(5x^3 - 4x^2 + 5) - (-2x^3 - 3x^2) =$

   A $3x^3 + x^2 + 5$
   B $3x^3 - 7x^2 + 5$
   C $7x^3 - x^2 + 5$
   D $7x^3 - 7x^2 + 5$

6. $(-z^3 - 4z^2 - 6) - (3z^3 - 6z + 5) =$

   A $-4z^3 - 4z^2 + 6z - 11$
   B $-2z^3 - 10z - 1$
   C $-4z^3 - 10z^2 - 1$
   D $-2z^2 + 2z - 11$

7. $(-7d^5)(-3d^2) =$

   A $-21d^7$
   B $21d^{10}$
   C $21d^7$
   D $-21d^{10}$

8. $(-5c^3d)(3c^5d^3)(2cd^4) =$

   A $30c^{15}d^8$
   B $15c^8d^{12}$
   C $-17c^{15}d^{12}$
   D $-30c^9d^8$

9. $-11j^2 \times -j^4 =$

   A $11j^6$
   B $11j^8$
   C $-11j^6$
   D $-11j^8$

10. $-6m^2(7m^2 + 5m - 6) =$

   A $-42m^2 + 30m^3 - 36$
   B $-42m^4 - 30m^3 + 36m^2$
   C $-13m^4 - m^2 + 36m^2$
   D $42m^4 - 30m^3 - 36m^2$

11. $-h^2(-4h + 5) =$

   A $-4h^3 - 5h^2$
   B $4h^3 - 5h^2$
   C $-5h^2 - 5h^2$
   D $-5h^3 - 5h^2$

12. $4m(m - 5) + 3m(2m^2 - 6m + 4) =$

   A $6m^3 - 14m^2 - 8m$
   B $-8m^2 - 8m - 1$
   C $7m - 14m^2 - 1$
   D $10m^2 - 26m - 20$

13. $2h(3h^2 - 5h - 2) + 4h(h^2 + 6h + 8) =$

   A $6h^3 + 19h^2 + 28h$
   B $-8m^2 - 8m - 1$
   C $7m - 14m^2 - 1$
   D $10h^3 + 14h^2 + 28h$

14. Multiply the following binomial and simplify. $(x - 3)(x + 3)$

   A $x^2 - 3x + 3x - 9$
   B $x^2 - 9$
   C $x^2 + 9$
   D $x^2 + 6x + 9$

15. Multiply the following binomial and simplify. $(x + 9)(x + 1)$

   A $x^2 + 10x + 9$
   B $x^2 + 10x + 10$
   C $x^2 + 9x + 9$
   D $x^2 + 9x + x + 9$

16. Multiply the following binomial and simplify. $(x - 2)^2$

   A $x^2 - 4x - 4$
   B $x^2 - 2x + 4$
   C $x^2 - 2x - 4$
   D $x^2 - 4x + 4$

17. $(x + 4)^2 = ?$

   A $x^2 + 4$
   B $x^2 + 16$
   C $x^2 + 16x + 8$
   D $x^2 + 8x + 16$

18. Simplify: $|4.2 - 5.1| + |-2.6|$

   A 1.7
   B -3.5
   C 3.5
   D -1.7

Chapter 7
Factoring

HSGE

Mathematics

This chapter covers the following Alabama objectives and standards in mathematics:

	Objective(s)
Standard I	4

7.1   Finding the Greatest Common Factor of Polynomials

In a multiplication problem, the numbers multiplied together are called **factors**. The answer to a multiplication problem is a called the **product**.

In the multiplication problem $5 \times 4 = 20$, 5 and 4 are factors and 20 is the product.

If we reverse the problem, $20 = 5 \times 4$, we say we have **factored** 20 into 5×4.

In this chapter, we will factor **polynomials**.

Example 1:   Find the greatest common factor of $2y^3 + 6y^2$.

   Step 1:   Look at the whole numbers. The greatest common factor of 2 and 6 is 2. Factor the 2 out of each term.

   $2\left(y^3 + 3y^2\right)$

   Step 2:   Look at the remaining terms, $y^3 + 3y^2$. What are the common factors of each term?

$$
\begin{array}{rcl}
y^3 & = & y \times \boxed{y \times y} \\
3y^2 & = & 3 \times \boxed{y \times y}
\end{array}
\longleftarrow \text{ common factors} = y^2
$$

   Step 3:   Factor 2 and y^2 out of each term: $2y^2(y+3)$

   Check: $2y^2(y+3) = 2y^3 + 6y^2$

Factor by finding the greatest common factor in each of the following.

1. $6x^4 + 18x^2$

2. $14y^3 + 7y$

3. $4b^5 + 12b^3$

4. $10a^3 + 5$

5. $2y^3 + 8y^2$

6. $6x^4 - 12x^2$

7. $18y^2 - 12y$

8. $15a^3 - 25a^2$

9. $4x^3 + 16x^2$

10. $6b^2 + 21b^5$

11. $27m^3 + 18m^4$

12. $100x^4 - 25x^3$

13. $4b^4 - 12b^3$

14. $18c^2 + 24c$

15. $20y^3 + 30y^5$

16. $16x^2 - 24x^5$

17. $15a^4 - 25a^2$

18. $24b^3 + 16b^6$

19. $36y^4 + 9y^2$

20. $42x^3 + 49x$

Factoring larger polynomials with 3 or 4 terms works the same way.

Example 2: $4x^5 + 16x^4 + 12x^3 + 8x^2$

Step 1:   Find the greatest common factor of the whole numbers. 4 can be divided evenly into 4, 16, 12, and 8; therefore, 4 is the greatest common factor.

Step 2:   Find the greatest common factor of the variables. x^5, x^4, x^3, and x^2 can be divided by x^2, the lowest power of x in each term.

$$4x^5 + 16x^4 + 12x^3 + 8x^2 = 4x^2 \left(x^3 + 4x^2 + 3x + 2\right)$$

Factor each of the following polynomials.

1. $5a^3 + 15a^2 + 20a$

2. $18y^4 + 6y^3 + 24y^2$

3. $12x^5 + 21x^3 + x^2$

4. $6b^4 + 3b^3 + 15b^2$

5. $14c^3 + 28c^2 + 7c$

6. $15b^4 - 5b^2 + 20b$

7. $t^3 + 3t^2 - 5t$

8. $8a^3 - 4a^2 + 12a$

9. $16b^5 - 12b^4 - 10b^2$

10. $20x^4 + 16x^3 - 24x^2 + 28x$

11. $40b^7 + 30b^5 - 50b^3$

12. $20y^4 - 15y^3 + 30y^2$

13. $4x^5 + 8x^4 + 12x^3 + 6x^2$

14. $16x^5 + 20x^4 - 12x^3 + 24x^2$

15. $18y^4 + 21y^3 - 9y^2$

16. $3n^5 + 9n^3 + 12n^2 + 15n$

17. $4d^6 - 8d^2 + 2d$

18. $10w^2 + 4w + 2$

19. $6t^3 - 3t^2 + 9t$

20. $25p^5 - 10p^3 - 5p^2$

21. $18x^4 + 9x^2 - 36x$

22. $6b^4 - 12b^2 - 6b$

23. $y^3 + 3y^2 - 9y$

24. $10x^5 - 2x^4 + 4x^2$

Example 3: Find the greatest common factor of $4a^3b^2 - 6a^2b^2 + 2a^4b^3$

Step 1: The greatest common factor of the whole numbers is 2.

$$4a^3b^2 - 6a^2b^2 + 2a^4b^3 = 2\left(2a^3b^2 - 3a^2b^2 + a^4b^3\right)$$

Step 2: Find the lowest power of each variable that is in each term. Factor them out of each term. The lowest power of a is a^2. The lowest power of b is b^2.

$$4a^3b^2 - 6a^2b^2 + 2a^4b^3 = 2a^2b^2\left(2a - 3 + a^2b\right)$$

Factor each of the following polynomials.

1. $3a^2b^2 - 6a^3b^4 + 9a^2b^3$

2. $12x^4y^3 + 18x^3y^4 - 24x^3y^3$

3. $20x^2y - 25x^3y^3$

4. $12x^2y - 20x^2y^2 + 16xy^2$

5. $8a^3b + 12a^2b + 20a^2b^3$

6. $36c^4 + 42c^3 + 24c^2 - 18c$

7. $14m^3n^4 - 28m^3n^2 + 42m^2n^3$

8. $16x^4y^2 - 24x^3y^2 + 12x^2y^2 - 8xy^2$

9. $32c^3d^4 - 56c^2d^3 + 64c^3d^2$

10. $21a^4b^3 + 27a^2b^3 + 15a^3b^2$

11. $4w^3t^2 + 6w^2t - 8wt^2$

12. $5pw^3 - 2p^2q^2 - 9p^3q$

13. $49x^3t^3 + 7xt^2 - 14xt^3$

14. $9cd^4 - 3d^4 - 6c^2d^3$

15. $12a^2b^3 - 14ab + 10ab^2$

16. $25x^4 + 10x - 20x^2$

17. $bx^3 - b^2x^2 + b^3x$

18. $4k^3a^2 + 22ka + 16k^2a^2$

19. $33w^4y^2 - 9w^3y^2 + 24w^2y^2$

20. $18x^3 - 9x^5 + 27x^2$

7.2   Finding the Numbers

The next kind of factoring we will do requires thinking of two numbers with a certain sum and a certain product.

Example 4:   Which two numbers have a sum of 8 and a product of 12? In other words, what pair of numbers would answer both equations?

$$\underline{\hspace{1cm}} + \underline{\hspace{1cm}} = 8 \quad \text{and} \quad \underline{\hspace{1cm}} \times \underline{\hspace{1cm}} = 12$$

You may think $4 + 4 = 8$, but 4×4 does not equal 12.
Or you may think $7 + 1 = 8$, but 7×1 does not equal 12.

$6 + 2 = 8$ and $6 \times 2 = 12$, so 6 and 2 are the pair of numbers that will work in both equations.

For each problem below, find one pair of numbers that will solve both equations.

1. $\underline{\hspace{1cm}} + \underline{\hspace{1cm}} = 14$ and $\underline{\hspace{1cm}} \times \underline{\hspace{1cm}} = 40$

2. $\underline{\hspace{1cm}} + \underline{\hspace{1cm}} = 10$ and $\underline{\hspace{1cm}} \times \underline{\hspace{1cm}} = 21$

3. $\underline{\hspace{1cm}} + \underline{\hspace{1cm}} = 18$ and $\underline{\hspace{1cm}} \times \underline{\hspace{1cm}} = 81$

4. $\underline{\hspace{1cm}} + \underline{\hspace{1cm}} = 12$ and $\underline{\hspace{1cm}} \times \underline{\hspace{1cm}} = 20$

5. $\underline{\hspace{1cm}} + \underline{\hspace{1cm}} = 7$ and $\underline{\hspace{1cm}} \times \underline{\hspace{1cm}} = 12$

6. $\underline{\hspace{1cm}} + \underline{\hspace{1cm}} = 8$ and $\underline{\hspace{1cm}} \times \underline{\hspace{1cm}} = 15$

7. $\underline{\hspace{1cm}} + \underline{\hspace{1cm}} = 10$ and $\underline{\hspace{1cm}} \times \underline{\hspace{1cm}} = 25$

8. $\underline{\hspace{1cm}} + \underline{\hspace{1cm}} = 14$ and $\underline{\hspace{1cm}} \times \underline{\hspace{1cm}} = 48$

9. $\underline{\hspace{1cm}} + \underline{\hspace{1cm}} = 12$ and $\underline{\hspace{1cm}} \times \underline{\hspace{1cm}} = 36$

10. $\underline{\hspace{1cm}} + \underline{\hspace{1cm}} = 17$ and $\underline{\hspace{1cm}} \times \underline{\hspace{1cm}} = 72$

11. $\underline{\hspace{1cm}} + \underline{\hspace{1cm}} = 15$ and $\underline{\hspace{1cm}} \times \underline{\hspace{1cm}} = 56$

12. $\underline{\hspace{1cm}} + \underline{\hspace{1cm}} = 9$ and $\underline{\hspace{1cm}} \times \underline{\hspace{1cm}} = 18$

13. $\underline{\hspace{1cm}} + \underline{\hspace{1cm}} = 13$ and $\underline{\hspace{1cm}} \times \underline{\hspace{1cm}} = 40$

14. $\underline{\hspace{1cm}} + \underline{\hspace{1cm}} = 16$ and $\underline{\hspace{1cm}} \times \underline{\hspace{1cm}} = 63$

15. $\underline{\hspace{1cm}} + \underline{\hspace{1cm}} = 10$ and $\underline{\hspace{1cm}} \times \underline{\hspace{1cm}} = 16$

16. $\underline{\hspace{1cm}} + \underline{\hspace{1cm}} = 8$ and $\underline{\hspace{1cm}} \times \underline{\hspace{1cm}} = 16$

17. $\underline{\hspace{1cm}} + \underline{\hspace{1cm}} = 9$ and $\underline{\hspace{1cm}} \times \underline{\hspace{1cm}} = 20$

18. $\underline{\hspace{1cm}} + \underline{\hspace{1cm}} = 13$ and $\underline{\hspace{1cm}} \times \underline{\hspace{1cm}} = 36$

19. $\underline{\hspace{1cm}} + \underline{\hspace{1cm}} = 15$ and $\underline{\hspace{1cm}} \times \underline{\hspace{1cm}} = 50$

20. $\underline{\hspace{1cm}} + \underline{\hspace{1cm}} = 11$ and $\underline{\hspace{1cm}} \times \underline{\hspace{1cm}} = 30$

7.3 More Finding the Numbers

Now that you have mastered positive numbers, take up the challenge of finding pairs of negative numbers or pairs where one number is negative and one is positive.

Example 5: Which two numbers have a sum of -3 and a product of -40? In other words, what pair of numbers would answer both equations?

$$\underline{\hspace{1cm}} + \underline{\hspace{1cm}} = -3 \quad \text{and} \quad \underline{\hspace{1cm}} \times \underline{\hspace{1cm}} = -40$$

It is faster to look at the factors of 40 first. 8 and 5 and 10 and 4 are possibilities. 8 and 5 have a difference of 3, and in fact, $5 + (-8) = -3$ and $5 \times (-8) = -40$. This pair of numbers, 5 and -8, will satisfy both equations.

For each problem below, find one pair of numbers that will solve both equations.

1. $\underline{\hspace{1cm}} + \underline{\hspace{1cm}} = -2$ and $\underline{\hspace{1cm}} \times \underline{\hspace{1cm}} = -35$
2. $\underline{\hspace{1cm}} + \underline{\hspace{1cm}} = 4$ and $\underline{\hspace{1cm}} \times \underline{\hspace{1cm}} = -5$
3. $\underline{\hspace{1cm}} + \underline{\hspace{1cm}} = 4$ and $\underline{\hspace{1cm}} \times \underline{\hspace{1cm}} = -12$
4. $\underline{\hspace{1cm}} + \underline{\hspace{1cm}} = -6$ and $\underline{\hspace{1cm}} \times \underline{\hspace{1cm}} = 8$
5. $\underline{\hspace{1cm}} + \underline{\hspace{1cm}} = 3$ and $\underline{\hspace{1cm}} \times \underline{\hspace{1cm}} = -40$
6. $\underline{\hspace{1cm}} + \underline{\hspace{1cm}} = 10$ and $\underline{\hspace{1cm}} \times \underline{\hspace{1cm}} = -11$
7. $\underline{\hspace{1cm}} + \underline{\hspace{1cm}} = 6$ and $\underline{\hspace{1cm}} \times \underline{\hspace{1cm}} = -27$
8. $\underline{\hspace{1cm}} + \underline{\hspace{1cm}} = 8$ and $\underline{\hspace{1cm}} \times \underline{\hspace{1cm}} = -20$
9. $\underline{\hspace{1cm}} + \underline{\hspace{1cm}} = -5$ and $\underline{\hspace{1cm}} \times \underline{\hspace{1cm}} = -24$
10. $\underline{\hspace{1cm}} + \underline{\hspace{1cm}} = -3$ and $\underline{\hspace{1cm}} \times \underline{\hspace{1cm}} = -28$
11. $\underline{\hspace{1cm}} + \underline{\hspace{1cm}} = -2$ and $\underline{\hspace{1cm}} \times \underline{\hspace{1cm}} = -48$
12. $\underline{\hspace{1cm}} + \underline{\hspace{1cm}} = -1$ and $\underline{\hspace{1cm}} \times \underline{\hspace{1cm}} = -20$
13. $\underline{\hspace{1cm}} + \underline{\hspace{1cm}} = -3$ and $\underline{\hspace{1cm}} \times \underline{\hspace{1cm}} = 2$
14. $\underline{\hspace{1cm}} + \underline{\hspace{1cm}} = 1$ and $\underline{\hspace{1cm}} \times \underline{\hspace{1cm}} = -30$
15. $\underline{\hspace{1cm}} + \underline{\hspace{1cm}} = -7$ and $\underline{\hspace{1cm}} \times \underline{\hspace{1cm}} = 12$
16. $\underline{\hspace{1cm}} + \underline{\hspace{1cm}} = 6$ and $\underline{\hspace{1cm}} \times \underline{\hspace{1cm}} = -16$
17. $\underline{\hspace{1cm}} + \underline{\hspace{1cm}} = 5$ and $\underline{\hspace{1cm}} \times \underline{\hspace{1cm}} = -24$
18. $\underline{\hspace{1cm}} + \underline{\hspace{1cm}} = -4$ and $\underline{\hspace{1cm}} \times \underline{\hspace{1cm}} = 4$
19. $\underline{\hspace{1cm}} + \underline{\hspace{1cm}} = -1$ and $\underline{\hspace{1cm}} \times \underline{\hspace{1cm}} = -42$
20. $\underline{\hspace{1cm}} + \underline{\hspace{1cm}} = -6$ and $\underline{\hspace{1cm}} \times \underline{\hspace{1cm}} = 8$

7.4   Factoring Trinomials

In the chapter on polynomials, you multiplied binomials (two terms) together, and the answer was a trinomial (three terms).

For example, $(x + 6)(x - 5) = x^2 + x - 30$

Now, you need to practice factoring a trinomial into two binomials.

Example 6:   Factor $x^2 + 6x + 8$

Step 1:   When the trinomial is in descending order as in the example above, you need to find a pair of numbers whose sum equals the number in the second term, while their product equals the third term. In the above example, find the pair of numbers that has a sum of 6 and a product of 8.

$$\underline{\hspace{1cm}} + \underline{\hspace{1cm}} = 6 \quad \text{and} \quad \underline{\hspace{1cm}} \times \underline{\hspace{1cm}} = 8$$

The pair of numbers that satisfy both equations is 4 and 2.

Step 2:   Use the pair of numbers in the binomials.

The factors of $x^2 + 6x + 8$ are $(x + 4)(x + 2)$

Check: To check, use the FOIL method.
$(x + 4)(x + 2) = x^2 + 4x + 2x + 8 = x^2 + 6x + 8$

Notice, when the second term and the third term of the trinomial are both positive, both numbers in the solution are positive.

Example 7:   Factor $x^2 - x - 6$ Find the pair of numbers where:

the sum is -1 and the product is -6

$$\underline{\hspace{1cm}} + \underline{\hspace{1cm}} = -1 \quad \text{and} \quad \underline{\hspace{1cm}} \times \underline{\hspace{1cm}} = -6$$

The pair of numbers that satisfies both equations is 2 and -3.
The factors of $x^2 - x - 6$ are $(x + 2)(x - 3)$

Notice, if the second term and the third term are negative, one number in the solution pair is positive, and the other number is negative.

Example 8: Factor $x^2 - 7x + 12$ Find the pair of numbers where:

the sum is -7 and the product is 12

_____ + _____ $= -7$ and   _____ $\times$ _____ $= 12$

The pair of numbers that satisfies both equations is -3 and -4
The factors of $x^2 - 7x + 12$ are $(x - 3)(x - 4)$.

Notice, if the second term of a trinomial is negative and the third term is positive, both numbers in the solution are negative.

Find the factors of the following trinomials.

1. $x^2 - x - 2$

2. $y^2 + y - 6$

3. $w^2 + 3w - 4$

4. $t^2 + 5t + 6$

5. $x^2 + 2x - 8$

6. $k^2 - 4k + 3$

7. $t^2 + 3t - 10$

8. $x^2 - 3x - 4$

9. $y^2 - 5y + 6$

10. $y^2 + y - 20$

11. $a^2 - a - 6$

12. $b^2 - 4b - 5$

13. $c^2 - 5c - 14$

14. $c^2 - c - 12$

15. $d^2 + d - 6$

16. $x^2 - 3x - 28$

17. $y^2 + 3y - 18$

18. $a^2 - 9a + 20$

19. $b^2 - 2b - 15$

20. $c^2 + 7c - 8$

21. $t^2 - 11t + 30$

22. $w^2 + 13w + 36$

23. $m^2 - 2m - 48$

24. $y^2 + 14y + 49$

25. $x^2 + 7x + 10$

26. $a^2 - 7a + 6$

27. $d^2 - 6d - 27$

7.5   More Factoring Trinomials

Sometimes a trinomial has a greatest common factor which must be factored out first.

Example 9:   Factor $4x^2 + 8x - 32$

Step 1:   Begin by factoring out the greatest common factor, 4.

$$4\left(x^2 + 2x - 8\right)$$

Step 2:   Factor by finding a pair of numbers whose sum is 2 and product is -8. 4 and -2 will work, so

$$4\left(x^2 + 2x - 8\right) = 4\left(x + 4\right)\left(x - 2\right)$$

Check: Multiply to check. $4\left(x + 4\right)\left(x - 2\right) = 4x^2 + 8x - 32$

Factor the following trinomials. Be sure to factor out the greatest common factor first.

1. $2x^2 + 6x + 4$

2. $3y^2 - 9y + 6$

3. $2a^2 + 2a - 12$

4. $4b^2 + 28b + 40$

5. $3y^2 - 6y - 9$

6. $10x^2 + 10x - 200$

7. $5c^2 - 10c - 40$

8. $6d^2 + 30d - 36$

9. $4x^2 + 8x - 60$

10. $6a^2 - 18a - 24$

11. $5b^2 + 40b + 75$

12. $3c^2 - 6c - 24$

13. $2x^2 - 18x + 28$

14. $4y^2 - 20y + 16$

15. $7a^2 - 7a - 42$

16. $6b^2 - 18b - 60$

17. $11d^2 + 66d + 88$

18. $3x^2 - 24x + 45$

7.6 Factoring More Trinomials

Some trinomials have a whole number in front of the first term that cannot be factored out of the trinomial. The trinomial can still be factored.

Example 10: Factor $2x^2 + 5x - 3$

Step 1: To get a product of $2x^2$, one factor must begin with $2x$ and the other with x.

$$(2x \quad)(x \quad)$$

Step 2: Now think: What two numbers give a product of -3? The two possibilities are 3 and -1 or -3 and 1. We know they could be in any order so there are 4 possible arrangements.

$$(2x + 3)(x - 1)$$
$$(2x - 3)(x + 1)$$
$$(2x + 1)(x - 3)$$
$$(2x - 1)(x + 3)$$

Step 3: Multiply each possible answer until you find the arrangement of the numbers that works. Multiply the outside terms and the inside terms and add them together to see which one will equal $5x$.

$$(2x + 3)(x - 1) = 2x^2 + x - 3$$
$$(2x - 3)(x + 1) = 2x^2 - x - 3$$
$$(2x + 1)(x - 3) = 2x^2 - 5 - 3$$
$$\boxed{(2x - 1)(x + 3) = 2x^2 + 5x - 3} \longleftarrow \text{ This arrangement works, therefore:}$$

The factors of $2x^2 + 5x - 3$ are $(2x - 1)(x + 3)$

Alternative: You can do some of the multiplying in your head. For the above example, ask yourself the following question: What two numbers give a product of -3 and give a sum of 5 (the whole number in the second term) when one number is first multiplied by 2 (the whole number in front of the first term)? The pair of numbers, -1 and 3, have a product of -3 and a sum of 5 when the 3 is first multiplied by 2. Therefore, the 3 will go in the opposite factor of the $2x$ so that when the terms are multiplied, you get -5.

You can use this method to at least narrow down the possible pairs of numbers when you have several from which to choose.

Factor the following trinomials.

1. $3y^2 + 14y + 8$

2. $5a^2 + 24a - 5$

3. $7b^2 + 30b + 8$

4. $2c^2 - 9c + 9$

5. $2y^2 - 7y - 15$

6. $3x^2 + 4x + 1$

7. $7y^2 + 13y - 2$

8. $11a^2 + 35a + 6$

9. $5y^2 + 17y - 12$

10. $3a^2 + 4a - 7$

11. $2a^2 + 3a - 20$

12. $5b^2 - 13b - 6$

13. $3y^2 - 4y - 32$

14. $2x^2 - 17x + 36$

15. $11x^2 - 29x - 12$

16. $5c^2 + 2c - 16$

17. $7y^2 - 30y + 27$

18. $2x^2 - 3x - 20$

19. $5b^2 - 19b - 4$

20. $7d^2 + 18d + 8$

21. $3x^2 - 20x + 25$

22. $2a^2 - 7a - 4$

23. $5m^2 + 12m + 4$

24. $9y^2 - 5y - 4$

25. $2b^2 - 13b + 18$

26. $7x^2 + 31x - 20$

27. $3c^2 - 2c - 21$

7.7   Factoring the Difference of Two Squares

Let's give an example of a **perfect square**.

25 is a perfect square because $5 \times 5 = 25$
49 is a perfect square because $7 \times 7 = 49$

Any variable with an even exponent is a perfect square.

y^2 is a perfect square because $y \times y = y^2$
y^4 is a perfect square because $y^2 \times y^2 = y^4$

When two terms that are both perfect squares are subtracted, factoring those terms is very easy. To factor the difference of perfect squares, you use the square root of each term, a plus sign in the first factor, and a minus sign in the second factor.

Example 11:   Factor $4x^2 - 9$

This example has two terms which are both perfect squares, and the terms are subtracted.

Step 1:   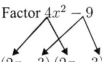 $(2x \quad 3)(2x \quad 3)$

Find the square root of each term.
Use the square roots in each of the factors.

Step 2:   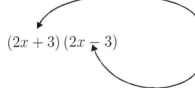 $(2x + 3)(2x - 3)$

Use a plus sign in one factor
and a minus sign in the other factor.

Check: Multiply to check. $(2x + 3)(2x - 3) = 4x^2 - 6x + 6x - 9 = 4x^2 - 9$

The inner and outer terms add to zero.

Example 12:   Factor $81y^4 - 1$

Step 1:   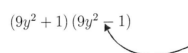 $(9y^2 + 1)(9y^2 - 1)$

Factor like the example above.
Notice, the second factor is also the difference of two perfect squares.

Step 2:   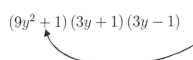 $(9y^2 + 1)(3y + 1)(3y - 1)$

Factor the second term further.
Note: You cannot factor the sum of two perfect squares.

Check: Multiply in reverse to check your answer.
$(9y^2 + 1)(3y + 1)(3y - 1) = (9y^2 + 1)(9y^2 - 3y + 3y - 1) =$
$(9y^2 + 1)(9y^2 - 1) = 81y^4 + 9y^2 - 9y^2 - 1 = 81y^4 - 1$

Factor the following differences of perfect squares.

1. $64x^2 - 49$

2. $4y^4 - 25$

3. $9a^4 - 4$

4. $25c^4 - 9$

5. $64y^2 - 9$

6. $x^4 - 16$

7. $49x^2 - 4$

8. $4d^2 - 25$

9. $9a^2 - 16$

10. $100y^4 - 49$

11. $c^4 - 36$

12. $36x^2 - 25$

13. $25x^2 - 4$

14. $9x^4 - 64$

15. $49x^2 - 100$

16. $16x^2 - 81$

17. $9y^4 - 1$

18. $49c^2 - 25$

19. $25d^2 - 64$

20. $36a^4 - 49$

21. $16x^4 - 16$

22. $b^2 - 25$

23. $c^4 - 144$

24. $9y^2 - 4$

25. $81x^4 - 16$

26. $4b^2 - 36$

27. $9w^2 - 9$

28. $64a^2 - 25$

29. $49y^2 - 121$

30. $x^6 - 9$

Chapter 7 Review

Factor the following polynomials completely.

1. $8x - 18$

2. $6x^2 - 18x$

3. $16b^3 + 8b$

4. $15a^3 + 40$

5. $20y^6 - 12y^4$

6. $5a - 15a^2$

7. $4y^2 - 36$

8. $2b^2 - 2b - 12$

9. $27y^2 + 42y - 5$

10. $12b^2 + 25b - 7$

11. $c^2 + cd - 20d^2$

12. $6y^2 + 30y + 36$

13. $2b^2 + 6b - 20$

14. $16b^4 - 81d^4$

15. $9w^2 - 54w - 63$

16. $12x^2 + 27x$

17. $2a^4 - 32$

18. $21c^2 + 41c + 10$

Chapter 7 Test

1. What is the greatest common factor of $4x^3$ and $8x^2$?

   A $4x^2$
   B $4x$
   C x^2
   D $8x$

2. Factor: $8x^4 - 7x^2 + 4x$

   A $4x\left(2x^3 - 7x + 4\right)$
   B $x\left(8x^4 - 7x^2 + 4x\right)$
   C $x\left(8x^3 - 7x + 4\right)$
   D $4x\left(2x^3 - 7x + 1\right)$

3. Factor: $x^2 + 6x + 8$

   A $(x + 2)(x + 4)$
   B $(x + 1)(x + 8)$
   C $(x - 2)(x - 4)$
   D $(x - 1)(x - 8)$

4. Factor: $2x^2 - 2x - 84$

   A $(2x + 7)(x - 12)$
   B $(2x - 12)(x + 7)$
   C $(2x - 7)(x + 12)$
   D $(2x + 12)(x - 7)$

5. Factor: $4x^2 - 64$

   A $(x - 8)(x + 8)$
   B $(4x - 8)(4x + 8)$
   C $(2x - 16)(2x + 16)$
   D $(2x - 8)(2x + 8)$

6. Factor the greatest common factor out of $2x^3 - 6x^2 + 2x$.

   A $2x\left(x^2 - 6x + 1\right)$
   B $2x\left(x^2 - 3x + 1\right)$
   C $2x\left(x^2 - 3x\right)$
   D $2\left(x^3 - 3x^2 + x\right)$

7. What are the factors of $x^2 + 10x + 25$?

   A $(x + 5)^2$
   B $(x - 5)^2$
   C $(x - 5)(x + 5)$
   D $(x + 5)(x + 2)$

8. What are the factors of $x^2 + 11x + 30$?

   A $(x + 6)(x + 5)$
   B $(x + 10)(x + 3)$
   C $(x + 6)^2$
   D $(x + 15)(x + 2)$

9. Factor: $3x^2 - 9$

   A $(3x + 3)(x - 3)$
   B $(x + 3)(x - 3)$
   C $3\left(x^2 - 3\right)$
   D $(3x + 9)(x - 1)$

Chapter 8
Solving Quadratic Equations

HSGE

Mathematics

This chapter covers the following Alabama objectives and standards in mathematics:

	Objective(s)
Standard II	2

In the previous chapter, we factored polynomials such as $y^2 - 4y - 5$ into two factors:

$$y^2 - 4y - 5 = (y + 1)(y - 5)$$

In this chapter, we learn that any equation that can be put in the form $ax^2 + bx + c = 0$ is a quadratic equation if a, b, and c are real numbers and $a \neq 0$. $ax^2 + bx + c = 0$ is the standard form of a quadratic equation. To solve these equations, follow the steps below.

Example 1: Solve $y^2 - 4y - 5 = 0$

Step 1: Factor the left side of the equation.

$$\begin{aligned} y^2 - 4y - 5 &= 0 \\ (y + 1)(y - 5) &= 0 \end{aligned}$$

Step 2: If the product of these two factors equals zero, then the two factors individually must be equal to zero. Therefore, to solve, we set each factor equal to zero.

$$\begin{array}{ll} (y + 1) = 0 & (y - 5) = 0 \\ \underline{-1 \quad -1} & \underline{+5 \quad +5} \\ y = -1 & y = 5 \end{array}$$

The equation has two solutions: $y = -1$ and $y = 5$

Check: To check, substitute each solution into the original equation.

When $y = -1$, the equation becomes:
$$\begin{aligned} (-1)^2 - (4)(-1) - 5 &= 0 \\ 1 + 4 - 5 &= 0 \\ 0 &= 0 \end{aligned}$$

When $y = 5$, the equation becomes:
$$\begin{aligned} 5^2 - (4)(5) - 5 &= 0 \\ 25 - 20 - 5 &= 0 \\ 0 &= 0 \end{aligned}$$

Both solutions produce true statements.
The solution set for the equation is $\{-1, 5\}$.

Solve each of the following quadratic equations by factoring and setting each factor equal to zero. Check by substituting answers back in the original equation.

1. $x^2 + x - 6 = 0$

2. $y^2 - 2y - 8 = 0$

3. $a^2 + 2a - 15 = 0$

4. $y^2 - 5y + 4 = 0$

5. $b^2 - 9b + 14 = 0$

6. $x^2 - 3x - 4 = 0$

7. $y^2 + y - 20 = 0$

8. $d^2 + 6d + 8 = 0$

9. $y^2 - 7y + 12 = 0$

10. $x^2 - 3x - 28 = 0$

11. $a^2 - 5a + 6 = 0$

12. $b^2 + 3b - 10 = 0$

13. $a^2 + 7a - 8 = 0$

14. $c^2 + 3x + 2 = 0$

15. $x^2 - x - 42 = 0$

16. $a^2 + a - 6 = 0$

17. $b^2 + 7b + 12 = 0$

18. $y^2 + 2y - 15 = 0$

19. $a^2 - 3a - 10 = 0$

20. $d^2 + 10d + 16 = 0$

21. $x^2 - 4x - 12 = 0$

Quadratic equations that have a whole number and a variable in the first term are solved the same way as the previous page. Factor the trinomial, and set each factor equal to zero to find the solution set.

Example 2:   Solve $2x^2 + 3x - 2 = 0$
$(2x - 1)(x + 2) = 0$
Set each factor equal to zero and solve:

$$
\begin{array}{rl}
2x - 1 & = 0 \\
+1 \quad +1 & \\
\hline
\dfrac{2x}{2} & = \dfrac{1}{2} \\
x & = \dfrac{1}{2}
\end{array}
\qquad
\begin{array}{rl}
x + 2 & = 0 \\
-2 \quad -2 & \\
\hline
x & = -2
\end{array}
$$

The solution set is $\left\{ \dfrac{1}{2}, -2 \right\}$.

Solve the following quadratic equations.

22. $3y^2 + 4y - 32 = 0$

23. $5c^2 - 2c - 16 = 0$

24. $7d^2 + 18d + 8 = 0$

25. $3a^2 - 10a - 8 = 0$

26. $11x^2 - 31x - 6 = 0$

27. $5b^2 + 17b + 6 = 0$

28. $3x^2 - 11x - 20 = 0$

29. $5a^2 + 47a - 30 = 0$

30. $2c^2 - 5c - 25 = 0$

31. $2y^2 + 11y - 21 = 0$

32. $5a^2 + 23a - 42 = 0$

33. $3d^2 + 11d - 20 = 0$

34. $3x^2 - 10x + 8 = 0$

35. $7b^2 + 23b - 20 = 0$

36. $9a^2 - 58a + 24 = 0$

37. $4c^2 - 25c - 21 = 0$

38. $8d^2 + 53d + 30 = 0$

39. $4y^2 - 29y + 30 = 0$

40. $8a^2 + 37a - 15 = 0$

41. $3x^2 - 41x + 26 = 0$

42. $8b^2 + 2b - 3 = 0$

8.1 Solving the Difference of Two Squares

To solve the difference of two squares, first factor. Then set each factor equal to zero.

Example 3: $25x^2 - 36 = 0$

Step 1:   Factor the left side of the equation.

$$25x^2 - 36 = 0$$
$$(5x + 6)(5x - 6) = 0$$

Step 2:   Set each factor equal to zero and solve.

$$
\begin{array}{c}
5x + 6 = 0 \\
\underline{\;\;-6 \quad -6\;\;} \\
\dfrac{5x}{5} = \dfrac{6}{5} \\
x = -\dfrac{6}{5}
\end{array}
\qquad
\begin{array}{c}
5x - 6 = 0 \\
\underline{\;\;+6 \quad +6\;\;} \\
\dfrac{5x}{5} = \dfrac{6}{5} \\
x = \dfrac{6}{5}
\end{array}
$$

Check: Substitute each solution in the equation to check.

for $x = -\dfrac{6}{5}$:

$$25x^2 - 36 = 0$$

$25\left(-\dfrac{6}{5}\right)\left(-\dfrac{6}{5}\right) - 36 = 0 \longleftarrow$ Substitute $-\frac{6}{5}$ for x.

$25\left(\dfrac{36}{25}\right) - 36 = 0 \longleftarrow$ Cancel the 25's.

$36 - 36 = 0 \longleftarrow$ A true statement. $x = -\frac{6}{5}$ is a solution.

for $x = \dfrac{6}{5}$:

$$25x^2 - 36 = 0$$

$25\left(\dfrac{6}{5}\right)\left(\dfrac{6}{5}\right) - 36 = 0 \longleftarrow$ Substitute $\frac{6}{5}$ for x.

$25\left(\dfrac{36}{25}\right) - 36 = 0 \longleftarrow$ Cancel the 25's.

$36 - 36 = 0 \longleftarrow$ A true statement. $x = \frac{6}{5}$ is a solution.

The solution set is $\left\{-\dfrac{6}{5}, \dfrac{6}{5}\right\}$.

Find the solution sets for the following.

1. $25a^2 - 16 = 0$

2. $c^2 - 36 = 0$

3. $9x^2 - 64 = 0$

4. $100y^2 - 49 - 0$

5. $4b^2 - 81 = 0$

6. $d^2 - 25 = 0$

7. $9x^2 - 1 = 0$

8. $16a^2 - 9 = 0$

9. $36y^2 - 1 = 0$

10. $36y^2 - 25 = 0$

11. $d^2 - 16 = 0$

12. $64b^2 - 9 = 0$

13. $81a^2 - 4 = 0$

14. $64y^2 - 25 = 0$

15. $4c^2 - 49 = 0$

16. $x^2 - 81 = 0$

17. $49b^2 - 9 = 0$

18. $a^2 - 64 = 0$

19. $9x^2 - 1 = 0$

20. $4y^2 - 9 = 0$

21. $t^2 - 100 = 0$

22. $16k^2 - 81 = 0$

23. $81a^2 - 4 = 0$

24. $36b^2 - 16 = 0$

8.2   Solving Perfect Squares

When the square root of a constant, variable, or polynomial results in a constant, variable, or polynomial without irrational numbers, the expression is a **perfect square**. Some examples are 49, x^2, and $(x-2)^2$.

Example 4:   Solve the perfect square for x. $(x-5)^2 = 0$

Step 1:   Take the square root of both sides.
$$\sqrt{(x-5)^2} = \sqrt{0}$$
$$(x-5) = 0$$

Step 2:   Solve the equation.
$$(x-5) = 0$$
$$x - 5 + 5 = 0 + 5$$
$$x = 5$$

Example 5:   Solve the perfect square for x. $(x-5)^2 = 64$

Step 1:   Take the square root of both sides.
$$\sqrt{(x-5)^2} = \sqrt{64}$$
$$(x-5) = \pm 8$$
$$(x-5) = 8 \text{ and } (x-5) = -8$$

Step 2:   Solve the two equations.

$(x-5) = 8$	and	$(x-5) = -8$
$x - 5 + 5 = 8 + 5$	and	$x - 5 + 5 = -8 + 5$
$x = 13$	and	$x = -3$

Solve the perfect square for x.

1. $(x-5)^2 = 0$

2. $(x+1)^2 = 0$

3. $(x+11)^2 = 0$

4. $(x-4)^2 = 0$

5. $(x-1)^2 = 0$

6. $(x+8)^2 = 0$

7. $(x+3)^2 = 4$

8. $(x-5)^2 = 16$

9. $(x-10)^2 = 100$

10. $(x+9)^2 = 9$

11. $(x-4.5)^2 = 25$

12. $(x+7)^2 = 36$

13. $(x+2)^2 = 49$

14. $(x-1)^2 = 4$

15. $(x+8.9)^2 = 49$

16. $(x-6)^2 = 81$

17. $(x-12)^2 = 121$

18. $(x+2.5)^2 = 64$

8.3   Using the Quadratic Formula

You may be asked to use the quadratic formula to solve an algebra problem known as a **quadratic equation**. The equation should be in the form $ax^2 + bx + c = 0$.

Example 6:   Using the quadratic formula, find x in the following equation: $x^2 - 8x = -7$.

Step 1:   Make sure the equation is set equal to 0.

$$x^2 - 8x + 7 = -7 + 7$$
$$x^2 - 8x + 7 = 0$$

The quadratic formula, $\dfrac{-b \pm \sqrt{b^2 - 4ac}}{2a}$, will be given to you on your formula sheet with your test.

Step 2:   In the formula, a is the number x^2 is multiplied by, b is the number x is multiplied by and c is the last term of the equation. For the equation in the example, $x^2 - 8x + 7$, $a = 1$, $b - -8$, and $c - 7$. When we look at the formula we notice a $\pm$ sign. This means that there will be two solutions to the equation, one when we use the plus sign and one when we use the minus sign. Substituting the numbers from the problem into the formula, we have:

$$\frac{8 + \sqrt{(-8)^2 - (4)(1)(7)}}{2(1)} = 7 \quad \text{or} \quad \frac{8 - \sqrt{(-8)^2 - (4)(1)(7)}}{2(1)} = 1$$

The solutions are $\{1, 7\}$.

For each of the following equations, use the quadratic formula to find two solutions.

1. $x^2 + x - 6 = 0$

2. $y^2 - 2y - 8 = 0$

3. $a^2 + 2a - 15 = 0$

4. $y^2 - 5y + 4 = 0$

5. $b^2 - 9b + 14 = 0$

6. $x^2 - 3x - 4 = 0$

7. $y^2 + y - 20 = 0$

8. $d^2 + 6d + 8 = 0$

9. $y^2 - 7y + 12 = 0$

10. $x^2 - 3x - 28 = 0$

11. $a^2 - 5a + 6 = 0$

12. $b^2 + 3b - 10 = 0$

13. $a^2 + 7a - 8 = 0$

14. $c^2 + 3c + 2 = 0$

15. $x^2 - x - 42 = 0$

16. $a^2 + 5a - 6 = 0$

17. $b^2 + 7b + 12 = 0$

18. $y^2 + y - 12 = 0$

19. $a^2 - 3a - 10 = 0$

20. $d^2 + 10d + 16 = 0$

21. $x^2 - 4x - 12 = 0$

Chapter 8 Review

Factor and solve each of the following quadratic equations.

1. $16b^2 - 25 = 0$

2. $a^2 - a - 30 = 0$

3. $x^2 - x = 6$

4. $100x^2 - 49 = 0$

5. $81y^2 = 9$

6. $y^2 = 21 - 4y$

7. $y^2 - 7y + 8 = 16$

8. $6x^2 + x - 2 = 0$

9. $3y^2 + y - 2 = 0$

10. $b^2 + 2b - 8 = 0$

11. $4x^2 + 19x - 5 = 0$

12. $8x^2 = 6x + 2$

13. $2y^2 - 6y - 20 = 0$

14. $-6x^2 + 7x - 2 = 0$

15. $y^2 + 3y - 18 = 0$

Using the quadratic formula, find both solutions for the variable.

16. $x^2 + 10x - 11 = 0$

17. $y^2 - 14y + 40 = 0$

18. $b^2 + 9b + 18 = 0$

19. $y^2 - 12y - 13 = 0$

20. $a^2 - 8a - 48 = 0$

21. $x^2 + 2x - 63 = 0$

Chapter 8 Test

1. Solve: $4y^2 - 9y = -5$

    A $\left\{1, \dfrac{5}{4}\right\}$

    B $\left\{-\dfrac{3}{4}, -1\right\}$

    C $\left\{-1, \dfrac{4}{5}\right\}$

    D $\left\{\dfrac{5}{16}, 1\right\}$

2. Solve for y: $2y^2 + 13y + 15 = 0$

    A $\left\{\dfrac{3}{2}, \dfrac{5}{2}\right\}$

    B $\left\{\dfrac{2}{3}, \dfrac{2}{5}\right\}$

    C $\left\{-5, -\dfrac{3}{2}\right\}$

    D $\left\{5, -\dfrac{3}{2}\right\}$

3. Solve for x.

    $x^2 - 3x - 18 = 0$

    A $\{-6, 3\}$
    B $\{6, -3\}$
    C $\{-9, 2\}$
    D $\{9, -2\}$

4. What are the values of x in the quadratic equation?

    $x^2 + 2x - 15 = x - 3$

    A $\{-4, 3\}$
    B $\{-3, 4\}$
    C $\{-3, 5\}$
    D Cannot be determined

5. Solve the equation $(x + 9)^2 = 49$

    A $x = -9, 9$
    B $x = -9, 7$
    C $x = -16, -2$
    D $x = -7, 7$

6. Solve the equation $c^2 + 8c - 9 = 0$ using the quadratic formula.

    A $c = \{1, -9\}$
    B $c = \{-1, 9\}$
    C $c = \{3, 3\}$
    D $c = \{-3, -3\}$

7. Solve $6a^2 + 11a - 10 = 0$, using the quadratic formula.

    A $\left\{-\dfrac{2}{5}, \dfrac{3}{2}\right\}$

    B $\left\{\dfrac{2}{5}, \dfrac{2}{3}\right\}$

    C $\left\{-\dfrac{5}{2}, \dfrac{2}{3}\right\}$

    D $\left\{\dfrac{5}{2}, \dfrac{2}{3}\right\}$

8. Solve: $x = \sqrt{\dfrac{10 - 8x}{2}}$

    A $\{0, 6\}$
    B $\{-5, 10\}$
    C $\{4, -2\}$
    D $\{-5, 1\}$

Chapter 9
Graphing and Writing Equations

This chapter covers the following Alabama objectives and standards in mathematics:

	Objective(s)
Standard IV	2
Standard V	1 & 4, 2
Standard VI	1

9.1   Graphing Linear Equations

In addition to graphing ordered pairs, the Cartesian plane can be used to graph the solution set for an equation. Any equation with two variables that are both to the first power is called a **linear equation.** The graph of a linear equation will always be a straight line.

Example 1:   Graph the solution set for $x + y = 7$.

Step 1:   Make a list of some pairs of numbers that will work in the equation.

$$\begin{array}{ll} \underline{x + y = 7} & \\ 4 + 3 = 7 & (4, 3) \\ -1 + 8 = 7 & (-1, 8) \\ 5 + 2 = 7 & (5, 2) \\ 0 + 7 = 7 & 0, 7 \end{array} \Bigg\} \text{ ordered pair solutions}$$

Step 2:   Plot these points on a Cartesian plane.

Step 3:   By passing a line through these points, we graph the solution set for $x + y = 7$. This means that every point on the line is a solution to the equation $x + y = 7$. For example, $(1, 6)$ is a solution, so the line passes through the point $(1, 6)$.

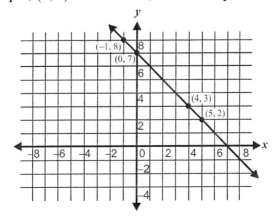

Make a table of solutions for each linear equation below. Then plot the ordered pair solutions on graph paper. Draw a line through the points. (If one of the points does not line up, you have made a mistake.)

1. $x + y = 6$

2. $y = x + 1$

3. $y = x - 2$

4. $x + 2 = y$

5. $x - 5 = y$

6. $x - y = 0$

Example 2:   Graph the equation $y = 2x - 5$.

Step 1:   This equation has 2 variables, both to the first power, so we know the graph will be a straight line. Substitute some numbers for x or y to find pairs of numbers that satisfy the equation. For the above equation, it will be easier to substitute values of x in order to find the corresponding value for y. Record the values for x and y in a table.

x	y
0	-5
1	-3
2	-1
3	1

If x is 0, y would be -5
If x is 1, y would be -3
If x is 2, y would be -1
If x is 3, y would be 1

Step 2:   Graph the ordered pairs, and draw a line through the points.

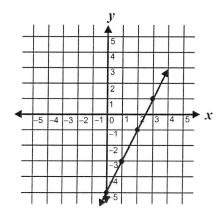

Find pairs of numbers that satisfy the equations below, and graph the line on graph paper.

1. $y = -2x + 2$

2. $2x - 2 = y$

3. $-x + 3 = y$

4. $y = x + 1$

5. $4x - 2 = y$

6. $y = 3x - 3$

7. $x = 4y - 3$

8. $2x = 3y + 1$

9. $x + 2y = 4$

9.2 Graphing Horizontal and Vertical Lines

The graph of some equations is a horizontal or a vertical line.

Example 3: $y = 3$

Step 1: Make a list of ordered pairs that satisfy the equation $y = 3$.

x	y
0	3
1	3
2	3
3	3

No matter what value of x you choose, y is always 3.

Step 2: Plot these points on an Cartesian plane, and draw a line through the points.

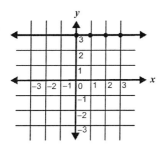

The graph is a horizontal line.

Example 4: $2x + 3 = 0$

Step 1: For these equations with only one variable, find what x equals first.
$$2x + 3 = 0$$
$$2x = -3$$
$$x = \frac{-3}{2}$$

Step 2: Using Example 3, find ordered pairs that satisfy the equation, plot the points, and graph the line.

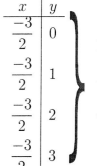

x	y
$\dfrac{-3}{2}$	0
$\dfrac{-3}{2}$	1
$\dfrac{-3}{2}$	2
$\dfrac{-3}{2}$	3

No matter which value of y you choose, the value of x does not change .

The graph is a vertical line.

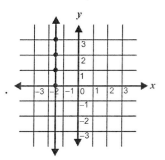

Find pairs of numbers that satisfy the equations below, and graph the line on graph paper.

1. $2y + 2 = 0$

2. $x = -4$

3. $3x = 3$

4. $y = 5$

5. $4x - 2 = 0$

6. $2x - 6 = 0$

7. $4y = 1$

8. $5x + 10 = 0$

9. $3y + 12 = 0$

10. $x + 1 = 0$

11. $2y - 8 = 0$

12. $3x = -9$

13. $x = -2$

14. $6y - 2 = 0$

15. $5x - 5 = 0$

9.3   Finding the Distance Between Two Points

Notice that a subscript added to the x and y identifies each ordered pair uniquely in the plane. For example, point 1 is identified as (x_1, y_1), point 2 as (x_2, y_2), and so on. This unique subscript identification allows us to calculate slope, distance, and midpoints of line segments in the plane using standard formulas like the distance formula. To find the distance between two points on a Cartesian plane, use the following formula:

$$d = \sqrt{(y_2 - y_1)^2 + (x_2 - x_1)^2}$$

Example 5:   Find the distance between $(-2, 1)$ and $(3, -4)$.

Plugging the values from the ordered pairs into the formula, we find:

$$d = \sqrt{(-4 - 1)^2 + [3 - (-2)]^2}$$

$$d = \sqrt{(-5)^2 + (5)^2}$$

$$d = \sqrt{25 + 25} = \sqrt{50}$$

To simplify, we look for perfect squares that are a factor of 50. $50 = 25 \times 2$. Therefore,

$$d = \sqrt{25} \times \sqrt{2} = 5\sqrt{2}$$

Find the distance between the following pairs of points using the distance formula above.

1. $(6, -1)\,(5, 2)$

2. $(-4, 3)\,(2, -1)$

3. $(10, 2)\,(6, -1)$

4. $(-2, 5)\,(-4, 3)$

5. $(8, -2)\,(3, -9)$

6. $(2, -2)\,(8, 1)$

7. $(3, 1)\,(5, 5)$

8. $(-2, -1)\,(3, 4)$

9. $(5, -3)\,(-1, -5)$

10. $(6, 5)\,(3, -4)$

11. $(-1, 0)\,(-9, -8)$

12. $(-2, 0)\,(-6, 6)$

13. $(2, 4)\,(8, 10)$

14. $(-10, -5)\,(2, -7)$

15. $(-3, 6)\,(1, -1)$

9.4 Finding the Midpoint of a Line Segment

You can use the coordinates of the endpoints of a line segment to find the coordinates of the midpoint of the line segment. The formula to find the midpoint between two coordinates is:

$$\text{midpoint, } M = \left(\frac{x_1 + x_2}{2}, \frac{y_1 + y_2}{2} \right)$$

Example 6: Find the midpoint of the line segment having endpoints at $(-3, -1)$ and $(4, 3)$.

Use the formula for the midpoint. $M = \left(\dfrac{4 + (-3)}{2}, \dfrac{3 + (-1)}{2} \right)$

When we simplify each coordinate, we find the midpoint, M, is $\left(\frac{1}{2}, 1 \right)$.

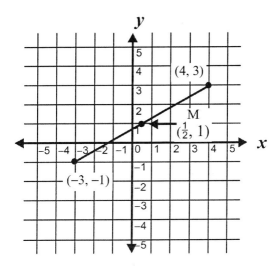

For each of the following pairs of points, find the coordinate of the midpoint, M, using the formula given above.

1. $(4, 5) \; (-6, 9)$

2. $(-3, 2) \; (-1, -2)$

3. $(3, 6) \; (9, 12)$

4. $(2, 5) \; (6, 9)$

5. $(8, 9) \; (6, 11)$

6. $(-4, 3) \; (8, 7)$

7. $(-1, -5) \; (-3, -11)$

8. $(4, 2) \; (-2, 8)$

9. $(4, 3) \; (-1, -5)$

10. $(-6, 2) \; (8, -8)$

11. $(-3, 9) \; (-9, 3)$

12. $(7, 8) \; (11, 6)$

13. $(12, 19) \; (2, 3)$

14. $(5, 4) \; (9, -2)$

15. $(-4, 6) \; (10, -2)$

9.5   Finding the Intercepts of a Line

The x-intercept is the point where the graph of a line crosses the x-axis. The y-intercept is the point where the graph of a line crosses the y-axis.

To find the x-intercept, set $y = 0$
To find the y-intercept, set $x = 0$

Example 7:   Find the x- and y-intercepts of the line $6x + 2y = 18$

   Step 1:   To find the x-intercept, set $y = 0$.

$$\begin{array}{rcl} 6x + 2\,(0) & = & 18 \\ 6x & = & 18 \\ \hline 6 & & 6 \\ x & = & 3 \end{array}$$

   The x-intercept is at the point $(3, 0)$.

   Step 2:   To find the y-intercept, set $x = 0$.

$$\begin{array}{rcl} 6\,(0) + 2y & = & 18 \\ 2y & = & 18 \\ \hline 2 & & 2 \\ y & = & 9 \end{array}$$

   The y-intercept is at the point $(0, 9)$.

   You can now use the two intercepts to graph the line.

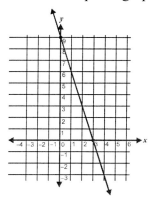

For each of the following equations, find both the x and the y intercepts of the line. For extra practice, draw each of the lines on graph paper.

1. $8x - 2y = 8$

2. $4x + 8y = 16$

3. $3x + 3y = 9$

4. $x - 2y = -5$

5. $8x + 4y = 32$

6. $3x - 4y = 12$

7. $-3x - 3y = 6$

8. $-6x + 2y = 18$

9. $4x - 2y = -4$

10. $-5x - 3y = 15$

11. $3x - 6y = -12$

12. $6x + 3y = 9$

13. $-2x - 6y = 18$

14. $2x + 3y = -6$

15. $-3x + 8y = 12$

9.6   Understanding Slope

The slope of a line refers to how steep a line is. Slope is also defined as the rate of change. When we graph a line using ordered pairs, we can easily determine the slope. Slope is often represented by the letter m.

$$\text{The formula for slope of a line is: } m = \frac{y_2 - y_1}{x_2 - x_1} \text{ or } \frac{\text{rise}}{\text{run}}$$

Example 8:   What is the slope of the following line that passes through the ordered pairs $(-4, -3)$ and $(1, 3)$?

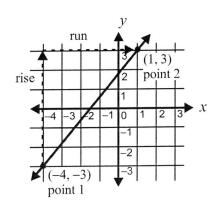

y_2 is 3, the y-coordinate of point 2.

y_1 is -3, the y-coordinate of point 1.

x_2 is 1, the x-coordinate of point 2.

x_1 is -4, the x-coordinate of point 1.

Use the formula for slope given above:

$$m = \frac{3 - (-3)}{1 - (-4)} = \frac{6}{5}$$

The slope is $\frac{6}{5}$. This shows us that we can go up 6 (rise) and over 5 to the right (run) to find another point on the line.

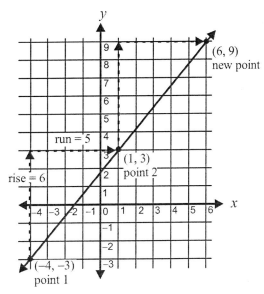

Example 9:   Find the slope of a line through the points $(-2, 3)$ and $(1, -2)$. It doesn't matter which pair we choose for point 1 and point 2. The answer is the same.

Let point 1 be $(-2, 3)$
Let point 2 be $(1, -2)$

$$\text{slope} = \frac{(y_2 - y_1)}{(x_2 - x_1)} = \frac{-2 - 3}{1 - (-2)} = \frac{-5}{3}$$

When the slope is negative, the line will slant left. For this example, the line will go **down** 5 units and then over 3 units to the **right**.

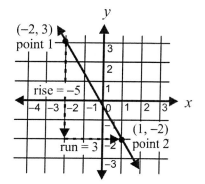

Example 10:   What is the slope of a line that passes through $(1, 1)$ and $(3, 1)$?

$$\text{slope} = \frac{1 - 1}{3 - 1} = \frac{0}{2} = 0$$

When $y_2 - y_1 = 0$, the slope will equal 0, and the line will be horizontal.

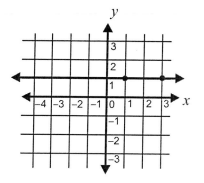

Example 11:   What is the slope of a line that passes through $(2, 1)$ and $(2, -3)$?

$$\text{slope} = \frac{-3 - 1}{2 - 2} = \frac{-4}{0} = \text{undefined}$$

When $x_2 - x_1 = 0$, the slope is undefined, and the line will be vertical.

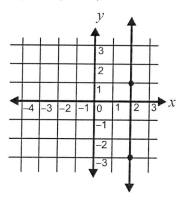

The following lines summarize what we know about slope.

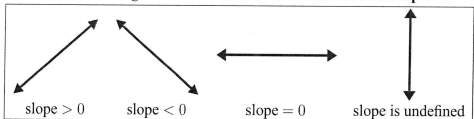

slope > 0 slope < 0 slope $= 0$ slope is undefined

Find the slope of the line that goes through the following pairs of points. Using graph paper, graph the line through the two points, and label the rise and run. (See Examples 8–11)

1. $(2, 3)$ $(4, 5)$

2. $(1, 3)$ $(2, 5)$

3. $(-1, 2)$ $(4, 1)$

4. $(1, -2)$ $(4, -2)$

5. $(3, 0)$ $(3, 4)$

6. $(3, 2)$ $(-1, 8)$

7. $(4, 3)$ $(2, 4)$

8. $(2, 2)$ $(1, 5)$

9. $(3, 4)$ $(1, 2)$

10. $(3, 2)$ $(3, 6)$

11. $(6, -2)$ $(3, -2)$

12. $(1, 2)$ $(3, 4)$

13. $(-2, 1)$ $(-4, 3)$

14. $(5, 2)$ $(4, -1)$

15. $(1, -3)$ $(-2, 4)$

16. $(2, -1)$ $(3, 5)$

9.7 Slope-Intercept Form of a Line

An equation that contains two variables, each to the first degree, is a **linear equation**. The graph for a linear equation is a straight line. To put a linear equation in slope-intercept form, solve the equation for y. This form of the equation shows the slope and the y-intercept. Slope-intercept form follows the pattern of $y = mx + b$. The "m" represents slope, and the "b" represents the y-intercept. The y-intercept is the point at which the line crosses the y-axis.

When the slope of a line is not 0, the graph of the equation shows a **direct variation** between y and x. When y increases, x increases in a certain proportion. The proportion stays constant. The constant is called the **slope** of the line.

Example 12: Put the equation $2x + 3y = 15$ in slope-intercept form. What is the slope of the line? What is the y-intercept? Graph the line.

Step 1: Solve for y:

$$\begin{array}{rcr} 2x + 3y &=& 15 \\ -2x && -2x \\ \hline \dfrac{3y}{3} &=& -\dfrac{2x}{3} + \dfrac{15}{3} \end{array}$$

slope-intercept form: $\quad y = -\frac{2}{3}x + 5$

The slope is $-\frac{2}{3}$ and the y-intercept is 5.

Step 2: Knowing the slope and the y-intercept, we can graph the line.

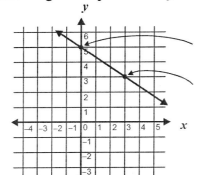

The y-intercept is 5, so the line passes through the point $(0, 5)$ on the y-axis.

The slope is $-\frac{2}{3}$, so go down 2 and over 3 to get a second point.

Put each of the following equations in slope-intercept form by solving for y. On your graph paper, graph the line using the slope and y-intercept.

1. $4x - 5y = 5$

6. $8x - 5y = 10$

11. $3x - 2y = -6$

16. $4x + 2y = 8$

2. $2x + 4y = 16$

7. $-2x + y = 4$

12. $3x + 4y = 2$

17. $6x - y = 4$

3. $3x - 2y = 10$

8. $-4x + 3y = 12$

13. $-x = 2 + 4y$

18. $-2x - 4y = 8$

4. $x + 3y = -12$

9. $-6x + 2y = 12$

14. $2x = 4y - 2$

19. $5x + 4y = 16$

5. $6x + 2y = 0$

10. $x - 5y = 5$

15. $6x - 3y = 9$

20. $6 = 2y - 3x$

9.8   Verify That a Point Lies on a Line

To know whether or not a point lies on a line, substitute the coordinates of the point into the formula for the line. If the point lies on the line, the equation will be true. If the point does not lie on the line, the equation will be false.

Example 13:   Does the point $(5, 2)$ lie on the line given by the equation $x + y = 7$?

Solution: Substitute 5 for x and 2 for y in the equation. $5 + 2 = 7$. Since this is a true statement, the point $(5, 2)$ does lie on the line $x + y = 7$.

Example 14:   Does the point $(0, 1)$ lie on the line given by the equation $5x + 4y = 16$?

Solution: Substitute 0 for x and 1 for y in the equation $5x + 4y = 16$. Does $5(0) + 4(1) = 16$? No, it equals 4, not 16. Therefore, the point $(0, 1)$ is not on the line given by the equation $5x + 4y = 16$.

For each point below, state whether or not it lies on the line given by the equation that follows the point coordinates.

1. $(2, 4)$ $6x - y = 8$

5. $(3, 7)$ $x - 5y = -32$

9. $(6, 8)$ $6x - y = 28$

2. $(1, 1)$ $6x - y = 5$

6. $(0, 5)$ $-6x - 5y = 3$

10. $(-2, 3)$ $x + 2y = 4$

3. $(3, 8)$ $-2x + y = 2$

7. $(2, 4)$ $4x + 2y = 16$

11. $(4, -1)$ $-x - 3y = -1$

4. $(9, 6)$ $-2x + y = 0$

8. $(9, 1)$ $3x - 2y = 29$

12. $(-1, -3)$ $2x + y = 1$

9.9   Graphing a Line Knowing a Point and Slope

If you are given a point of a line and the slope of a line, the line can be graphed.

Example 15:   Given that line l has a slope of $\frac{4}{3}$ and contains the point $(2, -1)$, graph the line.

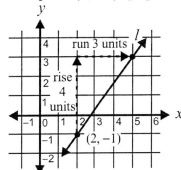

Plot and label the point $(2, -1)$
on a Cartesian plane.

The slope, m, is $\frac{4}{3}$, so the rise is
4, and the run is 3. From the point
$(2, -1)$, count 4 units up and
3 units to the right.

Draw the line through the two points.

Example 16:   Given a line that has a slope of $-\frac{1}{4}$ and passes through the point $(-3, 2)$, graph
the line.

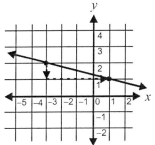

Plot the point $(-3, 2)$.

Since the slope is negative, go **down**
1 unit and over 4 units to get a second point.

Graph the line through the two points.

Graph a line on your own graph paper for each of the following problems. First, plot the point. Then use the slope to find a second point. Draw the line formed from the point and the slope.

1. $(2, -2), m = \frac{3}{4}$

2. $(3, -4), m = \frac{1}{2}$

3. $(1, 3), m = -\frac{1}{3}$

4. $(2, -4), m = 1$

5. $(3, 0), m = -\frac{1}{2}$

6. $(-2, 1), m = \frac{4}{3}$

7. $(-4, -2), m = \frac{1}{2}$

8. $(1, -4), m = \frac{3}{4}$

9. $(2, -1), m = -\frac{1}{2}$

10. $(5, -2), m = \frac{1}{4}$

11. $(-2, -3), m = \frac{2}{3}$

12. $(4, -1), m = -\frac{1}{3}$

13. $(-1, 5), m = \frac{2}{5}$

14. $(-2, 3), m = \frac{3}{4}$

15. $(4, 4), m = -\frac{1}{2}$

16. $(3, -3), m = -\frac{3}{4}$

17. $(-2, 5), m = \frac{1}{3}$

18. $(-2, -3), m = -\frac{3}{4}$

19. $(4, -3), m = \frac{2}{3}$

20. $(1, 4), m = -\frac{1}{2}$

9.10   Finding the Equation of a Line Using the Slope and Y-Intercept

Find the equation of a line is really simple if you are given two things: the slope and the y-intercept. First you need to know that the easiest equation to find when given these two things is the equation in slope-intercept form. Slope intercept form is $y = mx + b$, where m is the slope and b is the y-intercept.

Example 17:   Find the equation of a line given $m = \frac{1}{2}$ and y-intercept $= 4$.

$$y = mx + b \quad \longrightarrow \quad \text{Slope-Intercept form}$$

$$y = \tfrac{1}{2}x + 4 \quad \longrightarrow \quad \text{Plug in values for } m \text{ and } b.$$

Example 18:   Find the equation of a line that has a slope of 3 and a y-intercept of $-\frac{1}{3}$.

$$y = mx + b$$

$$y = 3x - \tfrac{1}{3}$$

Find the equation using the slope and y-intercept.

1. $m = 2$, y-intercept $= 4$

2. $m = 6$, y-intercept $= -2$

3. $m = -1$, y-intercept $= 7$

4. $m = -3$, y-intercept $= 0$

5. $m = -\frac{1}{2}$, y-intercept $= -11$

6. $m = \frac{2}{3}$, y-intercept $= 13$

7. $m = \frac{5}{4}$, y-intercept $= -\frac{4}{5}$

8. $m = -\frac{9}{2}$, y-intercept $= \frac{3}{9}$

9. $m = 1$, y-intercept $= -\frac{7}{5}$

10. $m = 0$, y-intercept $= 2$

11. $m = 2$, y-intercept $= -4$

12. $m = -7$, y-intercept $= 11$

13. $m = \frac{1}{2}$, y-intercept $= -5$

14. $m = 5$, y-intercept $= 6$

15. $m = -\frac{4}{5}$, y-intercept $= 8$

16. $m = 9$, y-intercept $= 1$

17. $m = \frac{3}{4}$, y-intercept $= 0$

18. $m = 7$, y-intercept $= -\frac{1}{4}$

19. $m = \frac{1}{7}$, y-intercept $= 12$

20. $m = \frac{6}{7}$, y-intercept $= \frac{7}{9}$

9.11 Finding the Equation of a Line Using Two Points or a Point and Slope

If you can find the slope of a line and know the coordinates of one point, you can write the equation for the line. You know the formula for the slope of a line is:

$$m = \frac{y_2 - y_1}{x_2 - x_1} \text{ or } \frac{y_2 - y_1}{x_2 - x_1} = m$$

Using algebra, you can see that if you multiply both sides of the equation by $x_2 - x_1$, you get:

$$y - y_1 = m(x - x_1) \longleftarrow \text{ point-slope form of an equation}$$

Example 19:   Write the equation of the line passing through the points $(-2, 3)$ and $(1, 5)$.

Step 1:   First, find the slope of the line using the two points given.
$$m = \frac{y_2 - y_1}{x_2 - x_1} = \frac{5 - 3}{1 - (-2)} = \frac{2}{3}$$

Step 2:   Pick one of the two points to use in the point-slope equation. For point $(-2, 3)$, we know $x_1 = -2$ and $y_1 = 3$, and we know $m = \frac{2}{3}$. Substitute these values into the point-slope form of the equation.

$$y - y_1 = m(x - x_1)$$

$$y - 3 = \frac{2}{3}[x - (-2)]$$

$$y - 3 = \frac{2}{3}x + \frac{4}{3}$$

$$y = \frac{2}{3}x + \frac{13}{3}$$

Use the point-slope formula to write an equation for each of the following lines.

1. $(1, -2), m = 2$
2. $(-3, 3), m = \frac{1}{3}$
3. $(4, 2), m = \frac{1}{4}$
4. $(5, 0), m = 1$
5. $(3, -4), m = \frac{1}{2}$

6. $(-1, -4)\ (2, -1)$
7. $(2, 1)\ (-1, -3)$
8. $(-2, 5)\ (-4, 3)$
9. $(-4, 3)\ (2, -1)$
10. $(3, 1)\ (5, 5)$

11. $(-3, 1), m = 2$
12. $(-1, 2), m = \frac{4}{3}$
13. $(2, -5), m = -2$
14. $(-1, 3), m = \frac{1}{3}$
15. $(0, -2), m = -\frac{3}{2}$

Chapter 9 Review

1. Graph the solution set for the linear equation: $x - 3 = y$.

2. Graph the equation $2x - 4 = 0$.

3. What is the slope of the line that passes through the points $(5, 3)$ and $(6, 1)$?

4. What is the slope of the line that passes through the points $(-1, 4)$ and $(-6, -2)$?

5. What is the x-intercept for the following equation? $6x - y = 30$

6. What is the y-intercept for the following equation? $4x + 2y = 28$

7. Graph the equation $3y = 9$.

8. Write the following equation in slope-intercept form.
$$3x = -2y + 4$$

9. What is the slope of the line $y = -\frac{1}{2}x + 3$?

10. What is the x-intercept of the line $y = 5x + 6$?

11. What is the y-intercept of the line $y - \frac{2}{3}x + 3 = 0$?

12. Graph the line which has a slope of -2 and a y-intercept of -3.

13. Find the equation of the line which contains the point $(0, 2)$ and has a slope of $\frac{3}{4}$.

14. What is the distance between the points $(3, 3)$ and $(6, -1)$?

15. What is the distance between the two points $(-3, 0)$ and $(2, 5)$?

16. Find the midpoint of the two points $(6, 10)$ and $(-4, 4)$.

17. Find the midpoint of the two points $(-1, -7)$ and $(5, 3)$.

Find the equation using the slope and y-intercept.

18. $m = 2$, y-intercept $= \frac{1}{4}$

19. $m = -4$, y-intercept $= -12$

20. $m = \frac{1}{2}$, y-intercept $= 17$

21. $m = -\frac{5}{2}$, y-intercept $= \frac{3}{2}$

Chapter 9 Test

1. Which is the graph of $x - 3y = 6$?

A

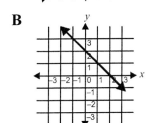

B

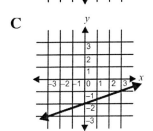

C

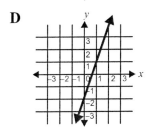

D

2. Which of the following points does not lie on the line $y = 3x - 2$?

  A $(0, -2)$
  B $(1, 1)$
  C $(-1, 5)$
  D $(2, 4)$

3. Which of the following is the graph of the equation $y = x - 3$?

A

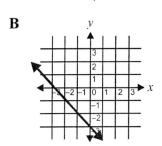

B

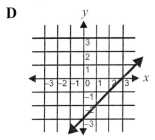

C

D

4. What is the x-intercept of the following linear equation? $3x + 4y = 12$

  A $(0, 3)$
  B $(3, 0)$
  C $(0, 4)$
  D $(4, 0)$

5. Which of the following equations is represented by the graph?

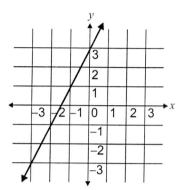

A $y = -3x + 3$

B $y = -\frac{1}{3}x + 3$

C $y = 3x - 3$

D $y = 2x + 3$

6. What is the equation of the line that includes the point $(4, -3)$ and has a slope of -2?

A $y = -2x - 5$

B $y = -2x - 2$

C $y = -2x + 5$

D $y = 2x - 5$

7. What is the x-intercept and y-intercept for the equation $x + 2y = 6$?

A x-intercept $= (0, 6)$
   y-intercept $= (3, 0)$

B x-intercept $= (4, 1)$
   y-intercept $= (2, 2)$

C x-intercept $= (0, 6)$
   y-intercept $= (0, 3)$

D x-intercept $= (6, 0)$
   y-intercept $= (0, 3)$

8. Which of the following graphs shows a line with a slope of 0 that passes through the point $(3, 2)$?

A

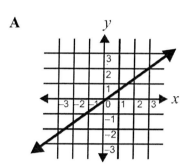

B

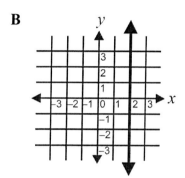

C

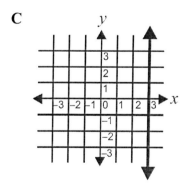

D

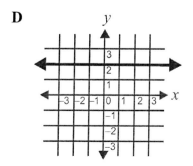

9. Put the following equation in slope-intercept form: $2x - 3y = 6$

A $y = \frac{2}{3}x - 2$

B $y = 2x - 2$

C $y = -\frac{2}{3}x + 2$

D $y = 2x + 2$

10. Look at the graphs below. Which of the following statements is false?

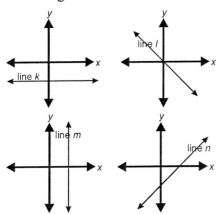

A The slope of line k is undefined.

B The slope of line l is negative.

C The slope of line m is undefined.

D The slope of line n is positive.

11. Which equation is represented by $m = -2$ and y-intercept $= 2$?

A $y = -2x - 2$

B $y = -2x + 2$

C $y = 2x + 2$

D $y = 2x - 2$

12. What is the y-intercept of $y = 2x - 6$?

A 6

B -6

C 2

D -2

13. Which equation passes through the points $(-1, 6)$ and $(0, 2)$?

A $y = 4x + 2$

B $y = -4x + 2$

C $y = 2x + 4$

D $y = -2x + 4$

14. Which of the following is not a solution of $3x = 5y - 1$?

A $(3, 2)$

B $(7, 4)$

C $\left(-\frac{1}{3}, 0\right)$

D $(-2, -1)$

15. $(-2, 1)$ is a solution for which of the following equations?

A $y + 2x = 4$

B $-2x - y = 5$

C $x + 2y = -4$

D $2x - y = -5$

Chapter 10
Systems of Equations

HSGE

Mathematics

This chapter covers the following Alabama objectives and standards in mathematics:

	Objective(s)
Standard II	3
Standard VII	8

Two linear equations considered at the same time are called a **system** of linear equations. The graph of a linear equation is a straight line. The graphs of two linear equations can show that the lines are **parallel, intersecting,** or **collinear.** Two lines that are **parallel** will never intersect and have no ordered pairs in common. If two lines are **intersecting,** they have one point in common, and in this chapter, you will learn to find the ordered pair for that point. If the graph of two linear equations is the same line, the lines are said to be **collinear.**

If you are given a system of two linear equations, and you put both equations in slope-intercept form, you can immediately tell if the graph of the lines will be **parallel, intersecting,** or **collinear.**

If two linear equations have the same slope and the same y-intercept, then they are both equations for the same line. They are called **collinear** or **coinciding** lines. A line is made up of an infinite number of points extending infinitely far in two directions. Therefore, collinear lines have an infinite number of points in common.

Example 1: $2x + 3y = -3$ **In slope intercept form:** $y = -\frac{2}{3}x - 1$

$4x + 6y = -6$ **In slope intercept form:** $y = -\frac{2}{3}x - 1$

The slope and y-intercept of both lines are the same.

If two linear equations have the same slope but different y-intercepts, they are **parallel** lines. Parallel lines never touch each other, so they have no points in common.

If two linear equations have different slopes, then they are intersecting lines and share exactly one point in common.

The chart below summarizes what we know about the graphs of two equations in slope-intercept form.

y-Intercepts	Slopes	Graphs	Number of Solutions
same	same	collinear	infinite
different	same	distinct parallel lines	none (they never touch)
same or different	different	intersecting lines	exactly one

For the systems of equations below, put each equation in slope-intercept form, and tell whether the graphs of the lines will be collinear, parallel, or intersecting.

1. $x - y = -1$
   $-x + y = -1$

2. $x - 2y = 4$
   $-x + 2y = 6$

3. $y - 2 = x$
   $x + 2 = y$

4. $x = y - 1$
   $-x = y - 1$

5. $2x + 5y = 10$
   $4x + 10y = 20$

6. $x + y = 3$
   $x - y = 1$

7. $2x = 4y - 6$
   $-6x + y = 3$

8. $x + y = 5$
   $2x + 2y = 10$

9. $2x = 3y - 6$
   $4x = 6y - 6$

10. $2x - 2y = 2$
    $3y = -x + 5$

11. $x = -y$
    $x = 4 - y$

12. $2x = y$
    $x + y = 3$

13. $x = y + 1$
    $y = x + 1$

14. $x - 2y = 4$
    $-2x + 4y = -8$

15. $2x + 3y = 4$
    $-2x + 3y = -8$

16. $2x - 4y = 1$
    $-6x + 12y = 3$

17. $-3x + 4y = 1$
    $6x + 8y = 2$

18. $x + y = 2$
    $5x + 5y = 10$

19. $x + y = 4$
    $x - y = 4$

20. $y = -x + 3$
    $x - y = 1$

10.1   Finding Common Solutions for Intersecting Lines

When two lines intersect, they share exactly one point in common.

Example 2: $3x + 4y = 20$ and $4x + 2y = 12$

Put each equation in slope-intercept form.

$$
\begin{array}{ll}
3x + 4y = 20 & 2y - 4x = 12 \\
4y = -3x + 20 & 2y = 4x + 12 \\
y = -\frac{3}{4}x + 5 & y = 2x + 6
\end{array}
$$

slope-intercept form

Straight lines with different slopes are **intersecting lines**. Look at the graphs of the lines on the same Cartesian plane.

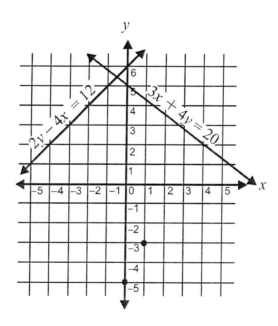

You can see from looking at the graph that the intersecting lines share one point in common. However, it is hard to tell from looking at the graph what the coordinates are for the point of intersection. To find the exact point of intersection, you can use the **substitution method** to solve the system of equations algebraically.

10.2   Solving Systems of Equations by Substitution

You can solve systems of equations by using the substitution method.

Example 3:   Find the point of intersection of the following two equations:

$$\text{Equation 1:} \quad x - y = 3$$
$$\text{Equation 2:} \quad 2x + y = 9$$

Step 1:   Solve one of the equations for x or y. Let's choose to solve equation 1 for x.

$$\text{Equation 1:} \quad x - y = 3$$
$$x = y + 3$$

Step 2:   Substitute the value of x from equation 1 in place of x in equation 2.

$$\text{Equation 2:} \quad 2x + y = 9$$
$$2(y + 3) + y = 9$$
$$2y + 6 + y = 9$$
$$3y + 6 = 9$$
$$3y = 3$$
$$y = 1$$

Step 3:   Substitute the solution for y back in equation 1 and solve for x.

$$\text{Equation 1:} \quad x - y = 3$$
$$x - 1 = 3$$
$$x = 4$$

Step 4:   The solution set is $(4, 1)$. Substitute in one or both of the equations to check.

$$\text{Equation 1:} \quad x - y = 3 \qquad \text{Equation 2:} \quad 2x + 9 = 9$$
$$4 - 1 = 3 \qquad\qquad\qquad\qquad 2(4) + 1 = 9$$
$$3 = 3 \qquad\qquad\qquad\qquad\quad 8 + 1 = 9$$
$$\qquad\qquad\qquad\qquad\qquad\qquad 9 = 9$$

The point $(4, 1)$ is common for both equations. This is the **point of intersection**.

For each of the following systems of equations, find the point of intersection, the common solution, using the substitution method.

1. $x + 2y = 8$
   $2x - 3y = 2$

2. $x - y = -5$
   $x + y = 1$

3. $x - y = 4$
   $x + y = 2$

4. $x - y = -1$
   $x + y = 9$

5. $-x + y = 2$
   $x + y = 8$

6. $x + 4y = 10$
   $x + 5y = 12$

7. $2x + 3y = 2$
   $4x - 9y = -1$

8. $x + 3y = 5$
   $x - y = 1$

9. $-x = y - 1$
   $x = y - 1$

10. $x - 2y = 2$
    $2y + x = -2$

11. $5x + 2y = 1$
    $2x + 4y = 10$

12. $3x - y = 2$
    $5x + y = 6$

13. $2x + 3y = 3$
    $4x + 5y = 5$

14. $x - y = 1$
    $-x - y = 1$

15. $x = y + 3$
    $y = 3 - x$

10.3   Solving Systems of Equations by Adding or Subtracting

You can solve systems of equations algebraically by adding or subtracting an equation from another equation or system of equations.

Example 4:   Find the point of intersection of the following two equations:
Equation 1: $x + y = 10$
Equation 2: $-x + 4y = 5$

Step 1:   Eliminate one of the variables by adding the two equations together. Since the x has the same coefficient in each equation, but opposite signs, it will cancel nicely by adding.

$$\begin{array}{r} x + y = 10 \\ + (-x + 4y = 5) \\ \hline 0 + 5y = 15 \\ 5y = 15 \\ y = 3 \end{array}$$

Add each like term together.
Simplify.
Divide both sides by 5.

Step 2:   Substitute the solution for y back into the equation, and solve for x.

Equation 1: $\quad x + y = 10 \qquad$ Substitute 3 for y.
$\qquad\qquad\quad x + 3 = 10 \qquad$ Subtract 3 from both sides.
$\qquad\qquad\qquad\quad x = 7$

Step 3:   The solution set is $(7, 3)$. To check, substitute the solution into both of the original equations.

Equation 1: $x + y = 10 \qquad\qquad$ Equation 2: $\qquad -x + 4y = 5$
$\qquad\qquad\quad 7 + 3 = 10 \qquad\qquad\qquad\qquad\qquad -(7) + 4(3) = 5$
$\qquad\qquad\qquad 10 = 10 \qquad\qquad\qquad\qquad\qquad\qquad -7 + 12 = 5$
$\qquad\qquad\qquad\qquad\qquad\qquad\qquad\qquad\qquad\qquad\qquad 5 = 5$

The point $(7, 3)$ is the point of intersection.

Example 5:   Find the point of intersection of the following two equations:
Equation 1: $3x - 2y = -1$
Equation 2: $-4y = -x - 7$

Step 1:   Put the variables on the same side of each equation. Take equation 2 out of y-intercept form.

$$\begin{array}{ll} -4y = -x - 7 & \text{Add } x \text{ to both sides.} \\ x - 4y = -x + x - 7 & \text{Simplify.} \\ x - 4y = -7 & \end{array}$$

Step 2:   Add the two equations together to cancel one variable. Since each variable has the same sign and different coefficients, we have to multiply one equation by a negative number so one of the variables will cancel. Equation 1's y variable has a coefficient of 2, and if multiplied by -2, the y will have the same variable as the y in equation 2, but a different sign. This will cancel nicely when added.

$$\begin{array}{ll} -2(3x - 2y = -1) & \text{Multiply by } -2. \\ -6x + 4y = 2 & \end{array}$$

Step 3:   Add the two equations.

$$-6x + 4y = 2$$
$$+ \underline{(x - 4y = -7)} \qquad \text{Add equation 2 to equation 1.}$$
$$-5x + 0 = -5 \qquad \text{Simplify.}$$
$$-5x = -5 \qquad \text{Divide both sides by } -5.$$
$$x = 1$$

Step 4:   Substitute the solution for x back into an equation and solve for y.

Equation 1:	$3x - 2y = -1$	Substitute 1 for x.
	$3(1) - 2y = -1$	Simplify.
	$3 - 2y = -1$	Subtract 3 from both sides.
	$3 - 3 - 2y = -1 - 3$	Simplify.
	$-2y = -4$	Divide both sides by -2.
	$y = 2$	

Step 5:   The solution set is $(1, 2)$. To check, substitute the solution into both of the original equations.

Equation 1: $\quad 3x - 2y = -1 \qquad$ Equation 2: $\quad -4y = -x - 7$
$3(1) - 2(2) = -1 \qquad\qquad\qquad -4(2) = -1 - 7$
$3 - 4 = -1 \qquad\qquad\qquad\qquad -8 = -8$
$-1 = -1$

The point $(1, 2)$ is the point of intersection.

For each of the following systems of equations, find the point of intersection by adding the 2 equations together. Remember you might need to change the coefficients and/or signs of the variables before adding.

1. $\begin{aligned} x + 2y &= 8 \\ -x - 3y &= 2 \end{aligned}$

2. $\begin{aligned} x - y &= 5 \\ 2x + y &= 1 \end{aligned}$

3. $\begin{aligned} x - y &= -1 \\ x + y &= 9 \end{aligned}$

4. $\begin{aligned} 3x - y &= -1 \\ x + y &= 13 \end{aligned}$

5. $\begin{aligned} -x + 4y &= 2 \\ x + y &= 8 \end{aligned}$

6. $\begin{aligned} x + 4y &= 10 \\ x + 7y &= 16 \end{aligned}$

7. $\begin{aligned} 2x - y &= 2 \\ 4x - 9y &= -3 \end{aligned}$

8. $\begin{aligned} x + 3y &= 13 \\ 5x - y &= 1 \end{aligned}$

9. $\begin{aligned} -x &= y - 1 \\ x &= y - 1 \end{aligned}$

10. $\begin{aligned} x - y &= 2 \\ 2y + x &= 5 \end{aligned}$

11. $\begin{aligned} 5x + 2y &= 1 \\ 4x + 8y &= 20 \end{aligned}$

12. $\begin{aligned} 3x - 2y &= 14 \\ x - y &= 6 \end{aligned}$

13. $\begin{aligned} 2x + 3y &= 3 \\ 3x + 5y &= 5 \end{aligned}$

14. $\begin{aligned} x - 4y &= 6 \\ -x - y &= -1 \end{aligned}$

15. $\begin{aligned} x &= 2y + 3 \\ y &= 3 - x \end{aligned}$

10.4   Solving Word Problems with Systems of Equations

Certain word problems can be solved using systems of equations.

Example 6:   In a game show, Andre earns 6 points for every right answer and loses 12 points for every wrong answer. He has answered correctly 12 times as many as he has missed. His final score was 120. How many times did he answer correctly?

Step 1:   Let r = number of right answers. Let w = number of wrong answers.

We know 2 sets of information that can be made into equations with 2 variables.

He earns $+6$ points for right answers and loses 12 points for wrong answers.

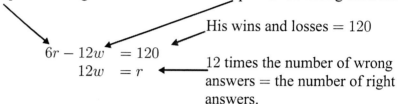

His wins and losses = 120

$$6r - 12w = 120$$
$$12w = r$$

12 times the number of wrong answers = the number of right answers.

Step 2:   Substitute the value for r ($12w$) in the first equation.

$$6(12w) - 12w = 120$$
$$w = 2$$

Step 3:   Substitute the value for w back in the equation.

$$6r - 12(2) = 120$$
$$r = 24$$

Example 7:   Ms. Sudberry bought pencils and stickers for her first grade class on two different days. The pencils and stickers cost the same each time she went to the store. How much did she pay for each pencil?

	Pencils	Stickers	Total Cost
Tuesday	30	40	$47.50
Saturday	60	5	$20.00

Step 1:   Set up your two equations. Let the price of pencils equal x, and the price of stickers equal y.

The amount of the pencils times the price of pencils (x) plus the amount of the stickers times the price of stickers (y) equals the total amount paid for both pencils and stickers.

Equation 1: $30x + 40y = \$47.50$
Equation 2: $60x + 5y = \$20.00$

Step 2:   Solve the equations by using one of the methods taught in this chapter. We will use the adding and subtracting method. First, multiply equation 1 by -2, so x will have the same coefficient in each equation but with opposite signs.
$-2(30x + 40y = \$47.50) = -60x - 80y = -\$95.00$

Step 3:   Add the new equation 1 to equation 2.

$$
\begin{array}{rrrrrr}
& -60x & - & 80y & = & -\$95.00 \\
+ & 60x & + & 5y & = & \$20.00 \\
\hline
& 0x & - & 75y & = & -\$75.00
\end{array}
$$

The new equation is $-75y = -\$75.00$.

Step 4:   Solve for y.
$-75y = -\$75.00$
$y = \$1.00$
Now, we know the price of stickers, but the question asked for the price of each pencil.

Step 5:   Substitute the value of y into either equation and solve for x to find the price of each pencil.
$30x + 40y = \$47.50$
$30x + 40\,(\$1.00) = \$47.50$
$30x + \$40.00 = \$47.50$
$30x = \$7.50$
$x = \$0.25$
The cost of each pencil is $\$0.25$.

Use systems of equations to solve the following word problems.

1. The sum of two numbers is 140 and their difference is 20. What are the two numbers?

2. The sum of two numbers is 126 and their difference is 42. What are the two numbers?

3. Kayla gets paid $6.00 for raking leaves and $8.00 for mowing the lawn of each of the neighbors in her subdivision. This year she mowed the lawns 12 times more than she raked leaves. In total, she made $918.00 for doing both. How many times did she rake the leaves?

4. Prices for the movie are $4.00 for children and $8.00 for adults. The total amount of ticket sales is $1,176. There are 172 tickets sold. How many adults and children buy tickets?

5. A farmer sells a dozen eggs at the market for $2.00 and one of his bags of grain for $5.00. He has sold 5 times as many bags of grain as he has dozens of eggs. By the end of the day, he has made $243.00 worth of sales. How many bags of grain did he sell?

6. Every time Lauren does one of her chores, she gets 15 minutes to talk on the phone. When she does not perform one of her chores, she gets 20 minutes of phone time taken away. This week she has done her chores 5 times more than she has not performed her chores. In total, she has accumulated 165 minutes. How many times has Lauren not performed her chores?

7. The choir sold boxes of candy and teddy bears near Valentine's Day to raise money. They sold twice as many boxes of candy as they did teddy bears. Bears sold for $8.00 each and candy sold for $6.00. They collected $380. How much of each item did they sell?

8. Mr. Marlow keeps ten and twenty dollar bills in his dresser drawer. He has 1 less than twice as many tens as twenties. He has $550 altogether. How many ten dollar bills does he have?

9. Kosta is a contestant on a math quiz show. For every correct answer, Kosta receives $18.00. For every incorrect answer, Kosta loses $24.00. Kosta answers the questions correctly twice as often as he answers the questions incorrectly. In total, Kosta wins $72.00. How many questions does Kosta answer incorrectly?

10. John Vasilovik works in landscaping. He gets paid $50 for each house he pressure-washes and $20 for each lawn he mows. He gets 4 times more jobs for mowing lawns than for pressure-washing houses. During a given month, John earns $2,600. How many houses does John pressure wash?

11. Every time Stephen walks the dog, he gets 30 minutes to play video or computer games. When he does not take out the dog on time, he gets a mess to clean up and loses 1 hour of video or computer game time. This week he has walked the dog on time 8 times more than he did not walk the dog on time. In total, he has accumulated 3 hours of video or computer time. How many times has Stephen not walked the dog on time?

12. On Friday, Rosa bought party hats and kazoos for her friend's birthday party. On Saturday she decided to purchase more when she found out more people were coming. How much did she pay for each party hat?

	Hats	Kazoos	Total Cost
Friday	15	20	$15.00
Saturday	10	5	$8.75

13. Timothy and Jesse went to purchase sports clothing they needed to play soccer. The table below shows what they bought and the amount they paid. What is the price of 1 soccer jersey?

	Soccer Jerseys	Tube Socks	Total Cost
Timothy	4	7	$78.30
Jesse	3	5	$57.60

Chapter 10 Review

For each system of equations below, tell whether the graphs of the lines will be collinear, parallel, or intersecting.

1. $y = 4x + 1$
   $y = 4x - 3$

2. $y - 4 = x$
   $2x + 8 = 2y$

3. $x + y = 5$
   $x - y = -1$

4. $2y - 3x = 6$
   $4y = 6x + 8$

5. $5y = 3x - 7$
   $4x - 3y = -7$

6. $2x - 2y = 2$
   $y - x = -1$

Find the common solution for each of the following systems of equations, using the substitution method.

7. $x - y = 2$
   $x + 4y = -3$

8. $x + y = 1$
   $x + 3y = 1$

9. $-4y = -2x + 4$
   $-x = -y - 3$

10. $2x + 8y = 20$
    $5y = 12 - x$

11. $x = y - 3$
    $-x = y + 3$

12. $-2x + y = -3$
    $x - y = 9$

Find the point of intersection for each system of equations by adding and/or subtracting the two equations.

13. $2x + y = 4$
    $3x - y = 6$

14. $x + 2y = 3$
    $x + 5y = 0$

15. $x + y = 1$
    $y = x + 7$

16. $2x + 4y = 5$
    $3x + 8y = 9$

17. $2x - 2y = 7$
    $3x - 5y = \frac{5}{2}$

18. $x - 3y = -2$
    $y = -\frac{1}{3}x + 4$

Use systems of equations to solve the following word problems.

19. Hargrove High School sold 227 tickets for their last basketball game. Adult tickets sold for \$5 and student tickets were \$2. How many adult tickets were sold if the ticket sales totalled \$574?

20. Zack is an ostrich and llama breeder. He sells full-grown ostriches for \$625 and full-grown llamas for \$750 each. Zack sold 1 less than 3 times as many llamas as ostriches this year. His total sales for the year were \$7,875.00. How many llamas did Zack sell during this year?

21. Sarah and Abdul played Geography Quiz Bowl during summer school. For every time Abdul got an answer right, Sarah got 4 answers right. If Sarah and Abdul correctly answered 75 questions, how many times did Abdul answer correctly?

Chapter 10 Test

1. Consider the following equations:

   $f(x) = 6x + 2$ and $f(x) = 3x + 2$

   Which of the following statements is true concerning the graphs of these equations?

   A The lines are collinear.
   B The lines intersect at exactly one point.
   C The lines are parallel to each other.
   D The graphs of the lines intersect each other at the point $(2, 2)$.

2. What is the solution to the following system of equations?

   $y = 4x - 8$
   $y = 2x$

   A $(-4, -8)$
   B $(4, 8)$
   C $(-1, -2)$
   D $(1, 2)$

3. Two lines are shown on the grid. One line passes through the origin and the other passes through $(-1, -1)$ with a y-intercept of 2. Which pair of equations below the grid identifies these lines?

   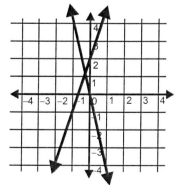

   A $y = \frac{1}{4}x$ and $y = \frac{1}{3}x + 2$
   B $x - 2y = 6$ and $4x + y = 4$
   C $y = 4x$ and $y = \frac{1}{3}x$
   D $y = 3x + 2$ and $y = -4x$

4. The graph of which system of equations below is parallel?

   A $x + 3y = 3$
          $3x + y = 3$
   B $3x + 3y = 6$
          $9x - 3y = 6$
   C $x - 3y = 6$
          $3y - x = -3$
   D $x + 3 = y$
          $x - 3 = 2y$

5. Which ordered pair is a solution for the following system of equations?

   $-3x + 7y = 25$
   $3x + 3y = -15$

   A $(-13, -2)$
   B $(-6, 1)$
   C $(-3, -2)$
   D $(-20, -5)$

6. For the following system of equations, find the point of intersection (common solution) using the substitution method.

   $-3x - y = -2$
   $5x + 2y = 20$

   A $(2, -4)$
   B $(2, 5)$
   C $(-16, 50)$
   D $\left(\frac{1}{5}, \frac{1}{2}\right)$

7. What is the intersection point of the graphs of the equations $2x - y = 2$ and $4x - 9y = -3$?

   A $\left(\frac{1}{2}, -1\right)$
   B $\left(1, \frac{3}{2}\right)$
   C $\left(-\frac{3}{2}, 1\right)$
   D $\left(\frac{3}{2}, 1\right)$

8. What is the intersection point of the graphs of the equations $x - 4y = 6$ and $-x - y = -1$?

   A $(2, 1)$
   B $(2, -1)$
   C $(-2, 1)$
   D $(-2, -1)$

9. The graphs of the equations $x = -y$ and $x = 4 - y$ are

   A collinear.
   B parallel.
   C intersecting.
   D not enough information.

10. The graphs of the equations $x + y = 2$ and $5x + 5y = 10$ are

    A collinear.
    B parallel.
    C intersecting.
    D not enough information.

11. What is the intersection point the graphs of the equations $x = y + 3$ and $y = 3 - x$?

    A $(3, 0)$
    B $(0, 3)$
    C $(3, 3)$
    D $(-3, 0)$

12. What is the intersection point of the graphs of the equations $2x + 3y = 2$ and $4x - 9y = -1$?

    A $(2, 3)$
    B $\left(\dfrac{1}{3}, \dfrac{1}{2} \right)$
    C $\left(\dfrac{1}{2}, \dfrac{1}{3} \right)$
    D $(3, 2)$

13. What is the intersection point of the graphs of the equations $-x = y - 1$ and $x = y - 1$?

    A $(0, 1)$
    B $(1, 0)$
    C $(-2, -1)$
    D $(2, 1)$

14. The admission fee at the fair is $1.50 for children and $4 for adults. On a certain day, 2,200 people enter the fair and $5,050 is collected. How many children attended? How many adults attended?

    A children = 750, adults = 1000
    B children = 1000, adults = 700
    C children = 1300, adults = 750
    D children = 1500, adults = 700

Chapter 11
Relations and Functions

This chapter covers the following Alabama objectives and standards in mathematics:

	Objective(s)
Standard III	1, 2
Standard V	1 & 4

11.1   Relations

A **relation** is a set of ordered pairs. We call the set of the first members of each ordered pair the **domain** of the relation. We call the set of the second members of each ordered pair the **range**.

Example 1:   State the domain and range of the following relation:
$$\{(2,4), (3,7), (4,9), (6,11)\}$$

Solution: Domain: $\{2, 3, 4, 6\}$ the first member of each ordered pair
    Range: $\{4, 7, 9, 11\}$ the second member of each ordered pair

State the domain and range for each relation.

1. $\{(2,5), (9,12), (3,8), (6,7)\}$

2. $\{(12,4), (3,4), (7,12), (26,19)\}$

3. $\{(4,3), (7,14), (16,34), (5,11)\}$

4. $\{(2,45), (33,43), (98,9), (43,61), (67,54)\}$

5. $\{(78,14), (29,67), (84,49), (16,18), (98,46)\}$

6. $\{(-8,16), (23,-7), (-4,-9), (16,-8), (-3,6)\}$

7. $\{(-7,-4), (-3,16), (-4,17), (-6,-8), (-8,12)\}$

8. $\{(-1,-2), (3,6), (-7,14), (-2,8), (-6,2)\}$

9. $\{(0,9), (-8,5), (3,12), (-8,-3), (7,18)\}$

10. $\{(58,14), (44,97), (74,32), (6,18), (63,44)\}$

11. $\{(-7,0), (-8,10), (-3,11), (-7,-32), (-2,57)\}$

12. $\{(18,34), (22,64), (94,36), (11,18), (91,45)\}$

When given an equation in two variables, the domain is the set of x values that satisfies the equation. The range is the set of y values that satisfies the equation.

Example 2: Find the range of the relation $3x = y + 2$ for the domain $\{-1, 0, 1, 2, 3\}$.
Solve the equation for each value of x given. The result, the y values, will be the range.

Given:

x	y
-1	
0	
1	
2	
3	

Solution:

x	y
-1	-5
0	-2
1	1
2	4
3	7

The range is $\{-5, -2, 1, 4, 7\}$.

Find the range of each relation for the given domain.

	Relation	**Domain**	**Range**
1.	$y = 5x$	$\{1, 2, 3, 4\}$	
2.	$y = |x|$	$\{-3, -2, -1, 0, 1\}$	
3.	$y = 3x + 2$	$\{0, 1, 3, 4\}$	
4.	$y = -|x|$	$\{-2, -1, 0, 1, 2\}$	
5.	$y = -2x + 1$	$\{0, 1, 3, 4\}$	
6.	$y = 10x - 2$	$\{-2, -1, 0, 1, 2\}$	
7.	$y = 3|x| + 1$	$\{-2, -1, 0, 1, 2\}$	
8.	$y - x = 0$	$\{1, 2, 3, 4\}$	
9.	$y - 2x = 0$	$\{1, 2, 3, 4\}$	
10.	$y = 3x - 1$	$\{0, 1, 3, 4\}$	
11.	$y = 4x + 2$	$\{0, 1, 3, 4\}$	
12.	$y = 2|x| - 1$	$\{-2 - 1, 0, 1, 2\}$	

11.2   Determining Domain and Range from Graphs

The domain is all of the x values that lie on the function in the graph from the lowest x value to the highest x value. The range is all of the y values that lie on the function in the graph from the lowest y to the highest y.

Example 3:   Find the domain and range of the graph.

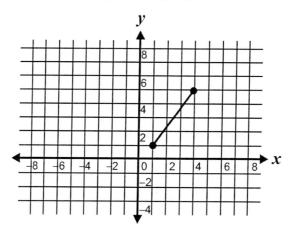

Step 1:   First find the lowest x value depicted on the graph. In this case it is 1. Then find the highest x value depicted on the graph. The highest value of x on the graph is 4. The domain must contain all of the values between the lowest x value and the highest x value. The easiest way to write this is $1 \leq \text{Domain} \leq 4$ or $1 \leq x \leq 4$.

Step 2:   Perform the same process for the range, but this time look at the lowest and highest y values. The answer is $1 \leq \text{Range} \leq 5$ or $1 \leq y \leq 5$.

Find the domain and range of each graph below. Write your answers in the line provided.

1.

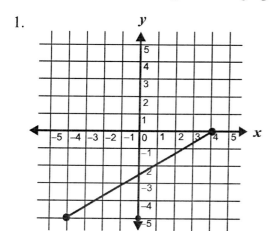

2.

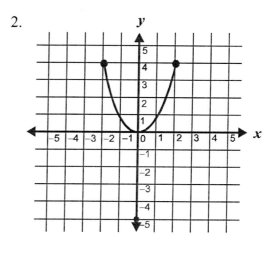

_____

_____

3.

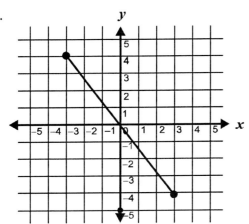

_____

4.

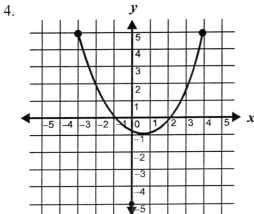

_____

5.

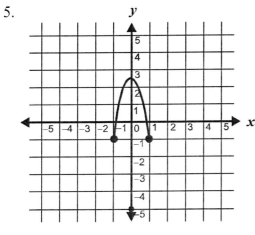

_____

6.

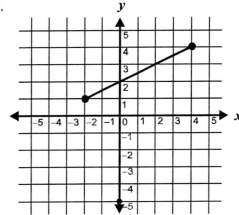

_____

7.

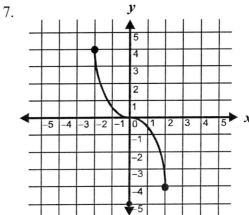

_____

8.

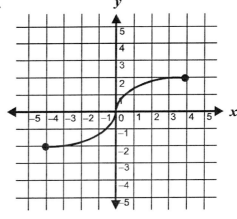

_____

11.3   Functions

Some relations are also **functions**. A relation is a function if **for every element in the domain, there is exactly one element in the range**. In other words, for each value for x there is only one unique value for y.

Example 4: $\{(2,4),(2,5),(3,4)\}$ is **NOT** a function because in the first pair, 2 is paired with 4, and in the second pair, 2 is paired with 5. The 2 can be paired with only one number to be a function. In this example, the x value of 2 has more than one value for y: 4 and 5.

Example 5: $\{(1,2),(3,2),(5,6)\}$ **IS** a function. Each first number is paired with only one second number. The 2 is repeated as a second number, but the relation remains a function.

Determine whether the ordered pairs of numbers below represent a function. Write "F" if it is a function. Write "NF" if it is not a function.

1. $\{(-1,1),(-3,3),(0,0),(2,2)\}$ _____

2. $\{(-4,-3),(-2,-3),(-1,-3),(2,-3)\}$ _____

3. $\{(5,-1),(2,0),(2,2),(5,3)\}$ _____

4. $\{(-3,3),(0,2),(1,1),(2,0)\}$ _____

5. $\{(-2,-5),(-2,-1),(-2,1),(-2,3)\}$ _____

6. $\{(0,2),(1,1),(2,2),(4,3)\}$ _____

7. $\{(4,2),(3,3),(2,2),(0,3)\}$ _____

8. $\{(-1,-1),(-2,-2),(3,-1),(3,2)\}$ _____

9. $\{(2,-2),(0,-2),(-2,0),(1,-3)\}$ _____

10. $\{(2,1),(3,2),(4,3),(5,-1)\}$ _____

11. $\{(-1,0),(2,1),(2,4),(-2,2)\}$ _____

12. $\{(1,4),(2,3),(0,2),(0,4)\}$ _____

13. $\{(0,0),(1,0),(2,0),(3,0)\}$ _____

14. $\{(-5,-1),(-3,-2),(-4,-9),(-7,-3)\}$ _____

15. $\{(8,-3),(-4,4),(8,0),(6,2)\}$ _____

16. $\{(7,-1),(4,3),(8,2),(2,8)\}$ _____

17. $\{(4,-3),(2,0),(5,3),(4,1)\}$ _____

18. $\{(2,-6),(7,3),(-3,4),(2,-3)\}$ _____

19. $\{(1,1),(3,-2),(4,16),(1,-5)\}$ _____

20. $\{(5,7),(3,8),(5,3),(6,9)\}$ _____

11.4   Function Notation

Function notation is used to represent relations which are functions. Some commonly used letters to represent functions include f, g, h, F, G, and H.

Example 6: $f(x) = 2x - 1$; find $f(-3)$

  Step 1:   Find $f(-3)$ means to replace x with -3 in the relation $2x - 1$.
  $f(-3) = 2(-3) - 1$

  Step 2:   Solve $f(-3)$. $f(-3) = 2(-3) - 1 = -6 - 1 = -7$
  $f(-3) = -7$

Example 7: $g(x) = 4 - 2x^2$; find $g(2)$

  Step 1:   Replace x with 2 in the relation $4 - 2x^2$.
  $g(2) = 4 - 2(2)^2$

  Step 2:   Solve $g(2)$. $g(2) = 4 - 2(2)^2 = 4 - 2(4) = 4 - 8 = -4$
  $g(2) = -4$

Find the solutions for each of the following.

1. $F(x) = 2 + 3x^2$; find $F(3)$

6. $G(x) = 4x^2 + 4$; find $G(0)$

2. $f(x) = 4x + 6$; find $f(-4)$

7. $f(x) = 7 - 6x$; find $f(-4)$

3. $H(x) = 6 - 2x^2$; find $H(-1)$

8. $h(x) = 2x^2 + 10$; find $h(5)$

4. $g(x) = -3x + 7$; find $g(-3)$

9. $F(x) = 7 - 5x$; find $F(2)$

5. $f(x) = -5 + 4x$; find $F(7)$

10. $f(x) = -4x^2 + 5$; find $f(-2)$

11.5   Graphs of Common Relations

A graph is an image that shows the relationship between two or more variables. In this section, we will learn how to graph six basic functions.

Example 8: $f(x) = x$

The notation $f(x)$ is the same as y, it just means f as a function of x

The easiest way to begin graphing this is to draw a table of values. To create an accurate graph, you should choose at least 5 values for your table. These values are now your (x, y) ordered pair and they are ready to be plotted.

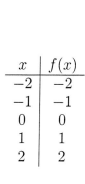

x	$f(x)$
-2	-2
-1	-1
0	0
1	1
2	2

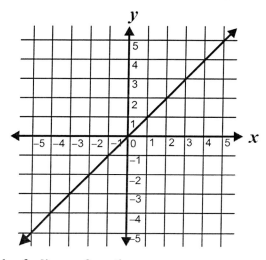

This is called the graph of a **linear function**.

Example 9: $f(x) = x^2$

Exponential functions can be graphed the same way as linear functions.

x	$f(x)$
-2	4
-1	1
0	0
1	1
2	4

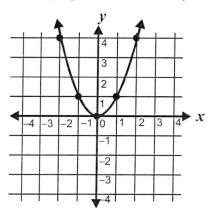

This is called the graph of a **quadratic function**.

Example 10: $f(x) = |x|$

Don't forget, when you take the absolute value of a negative number, it becomes positive!

x	$f(x)$
-2	2
-1	1
0	0
1	1
2	2

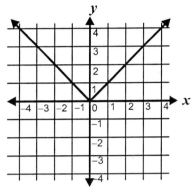

This is called the graph of an **absolute value function**.

Example 11: $f(x) = \sqrt{x}$

An equation like this will be easier to graph if you choose values of x that are perfect squares and because you can't take the square root of a negative, there is no need to select negative values.

x	$f(x)$
0	0
1	1
4	2
9	3

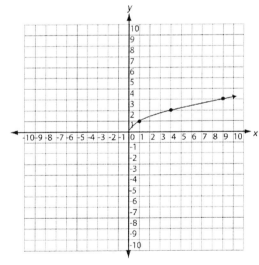

This is called the graph of a **square root function**.

Graph the following.

1. $f(x) = 2x$

2. $f(x) = -\left|\frac{1}{2}x\right|$

3. $f(x) = |4x|$

4. $f(x) = \sqrt{2x}$

5. $f(x) = \frac{1}{3}x^2$

6. $f(x) = -4x^2$

7. $f(x) = -\sqrt{x}$

8. $f(x) = -x$

9. $f(x) = 2x^2$

11.6   Recognizing Functions

Recall that a relation is a function with only one y value for every x value. We can depict functions in many ways including through graphs.

Example 12:

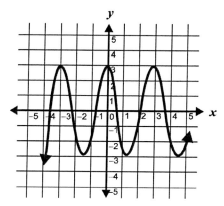

This graph **IS** a function because it has only one y value for each value of x.

Example 13:

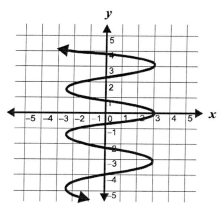

This graph is **NOT** a function because there is more than one y value for each value of x.

Hint: An easy way to determine a function from a graph is to do a vertical line test. First, draw a vertical line that crosses over the whole graph. If the line crosses the graph more than one time, then it is not a function. If it only crosses it once, it is a function. Take Example 11 above:

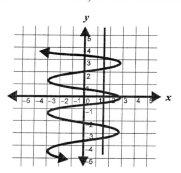

Since the vertical line passes over the graph more than one time, it is not a function.

Determine whether or not each of the following graphs is a function. If it is, write function on the line provided. If it is not a function, write NOT a function on the line provided.

1.
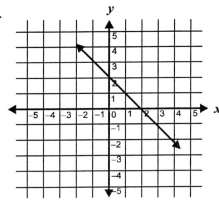

_____

4.
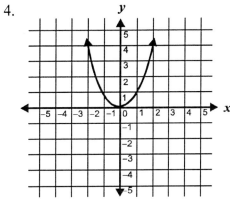

_____

2.
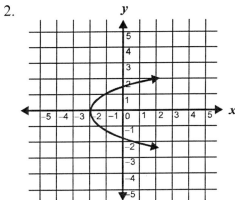

_____

5.
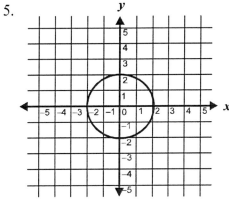

_____

3.
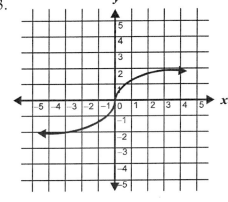

_____

6.
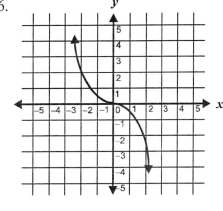

_____

7.

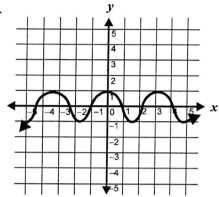

_____

8.

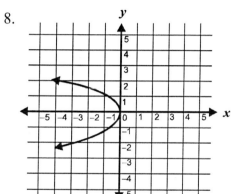

_____

9.

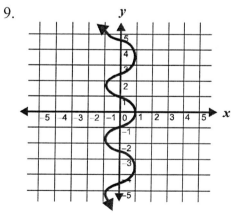

_____

10.

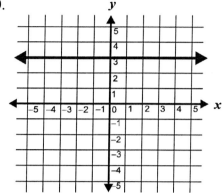

_____

11.

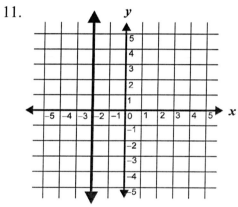

_____

12.

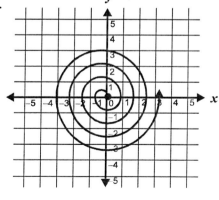

_____

11.7   Function Tables

Functions can also use a variable such as n to be the input of the function and $f(n)$, read "f of n" to represent the output of the function.

Example 14:   Function rule: $3n + 4$

n	$f(n)$
-1	1
0	4
1	7
2	10

Fill in the tables for each function rule below.

1.   rule: $2(n-5)$

n	$f(n)$
0	
1	
2	
3	

4.   rule: $2x(x-1)$

x	$f(x)$
1	
2	
3	
4	

7.   rule: $n(n+2)$

n	$f(n)$
1	
2	
3	
4	

2.   rule: $3x(x-4)$

x	$f(x)$
0	
1	
2	
3	

5.   rule: $\dfrac{1}{n+3}$

n	$f(n)$
1	
2	
3	
4	

8.   rule: $2x-3$

x	$f(x)$
1	
2	
3	
4	

3.   rule: $\dfrac{2-n}{2}$

n	$f(n)$
0	
2	
4	
6	
8	

6.   rule: $|4x|$

x	$f(x)$
-2	
-1	
0	
1	
2	

9.   rule: $|3-2n|$

n	$f(n)$
-2	
-1	
0	
1	
2	

11.8   Finding Equations from Tables

As you know, there are many different ways to find equations using other information. In this section, we will learn to determine equations using a table of values.

Example 15:

x	12	3	0	-6
y	4	1	0	-2

Use the chart to find the equation.

Step 1:   The first step when looking at a table is to look for the y-intercept or place where $x = 0$. According to the chart, $x = 0$ when $y = 0$, so when it is written in slope intercept form, no y-intercept will be shown. (i.e., $y = 2x$)

Step 2:   Find the slope. Pick two x values and their corresponding y values from the chart. Then, using the slope formula, substitute the chosen values into the formula to find the slope, m.

Slope Formula: $m = \dfrac{y_2 - y_1}{x_2 - x_1}$ points: $(3, 1)$ and $(0, 0)$

$$m = \frac{y_2 - y_1}{x_2 - x_1} = \frac{0 - 1}{0 - 3} = \frac{-1}{-3} = \frac{1}{3}$$

Step 3:   Now we know the slope and we also know that the y-intercept occurs at $x = 0$. Now we can write the equation. Substitute the values of the slope and y-intercept into the equation $y = mx + b$, where $m =$ slope and $b = y$-intercept.

$y = \frac{1}{3}x + 0 = \frac{1}{3}x$

Given the tables, find the equations.

1.

x	-2	-1	0	1	2
y	-1	1	3	5	7

2.

x	-2	-1	0	1	2
y	7	4	1	-2	-5

3.

x	-2	-1	0	1	2
y	3	$3\frac{1}{2}$	4	$4\frac{1}{2}$	5

4.

x	-2	-1	0	1	2
y	-10	-4	2	8	14

5.

x	-2	-1	0	1	2
y	-6	-5	-4	-3	-2

6.

x	-2	-1	0	1	2
y	$-\frac{1}{2}$	$\frac{1}{2}$	$1\frac{1}{2}$	$2\frac{1}{2}$	$3\frac{1}{2}$

7.

x	-2	-1	0	1	2
y	-6	-4	-2	0	2

8.

x	-2	-1	0	1	2
y	-39	-24	-9	6	21

9.

x	-2	-1	0	1	2
y	$-\frac{1}{6}$	$\frac{1}{6}$	$\frac{1}{2}$	$\frac{5}{6}$	$\frac{7}{6}$

10.

x	-2	-1	0	1	2
y	$-12\frac{1}{4}$	$-6\frac{1}{4}$	$-\frac{1}{4}$	$5\frac{3}{4}$	$11\frac{3}{4}$

Example 16:   Use the chart below to find the equation of the values.

x	-2	-1	1	2
y	-7	-4	2	5

Step 1:   Unfortunately, the y-intercept cannot always be found by looking at the chart. The y-intercept is not shown in the chart, so it must be found algebraically.

Step 2:   We will find the slope first. Use the points $(0, -1)$ and $(1, 2)$.

$$m = \frac{y_2 - y_1}{x_2 - x_1} = \frac{2 - (-1)}{1 - 0} = \frac{3}{1} = 3$$

The slope is 3.

Step 3:   After finding the slope, we can use the equation below by choosing a random point from the table and substituting it in to find the equation of the values in the chart.

$$y - y_1 = m(x - x_1)$$

Pick a random point from the table. We chose $(1, 2)$. Substitute the slope for m and the random point for x_1 and y_1 in the equation.

$$y - y_1 = m(x - x_1)$$
$$y - 2 = 3(x - 1)$$

Step 4:   Simplify.

$$y - 2 = 3x - 3$$
$$y = 3x - 1$$

Given the tables, find the equations.

1.

x	2	4	6	8	10
y	1	2	3	4	5

2.

x	1	2	3	4	5	6
y	0	2	4	6	8	10

3.

x	4	8	12	16	20	24	28
y	4	5	6	7	8	9	10

4.

x	2	4	6	8	10	12	14
y	0	1	2	3	4	5	6

5.

x	1	2	3	4	5	6	7
y	-2	1	4	7	10	13	16

6.

x	3	6	9	12	15	18	21
y	4	6	8	10	12	14	16

7.

x	3	6	9	12	15	18	21
y	1	5	9	13	17	21	25

10.

x	1	2	3	4	5	6
y	40	30	20	10	0	-10

8.

x	1	2	3	4	5	6
y	$-\frac{7}{2}$	-3	$-\frac{5}{2}$	-2	$-\frac{3}{2}$	-1

11.

x	1	2	3	4	5	6	7
y	3	1	-1	-3	-5	-7	-9

9.

x	2	3	4	5	6
y	6.50	9.50	12.50	15.50	18.50

12.

x	1	2	3	4	5
y	-24	-44	-64	-84	-104

Write an equation for each of the following charts.

13.

Cost for Hiring a Plumber

Number of Hours (x)	Total Charges (y)
1	$145
2	$230
3	$315
4	$400
5	$485
6	$570

16.

Charges for Dog Walking

Number of Minutes (x)	Total Charges (y)
1	$12.50
5	$14.50
10	$17.00
15	$19.50
20	$22.00
25	$24.50

14.

Carl's Carpet Cleaning Charges

Number of Rooms (x)	Total Charges (y)
1	$87
2	$99
3	$111
4	$123
5	$135
6	$147

17.

Dave's Carpet Cleaning Charges

Number of Rooms (x)	Total Charges (y)
1	$80
2	$95
3	$110
4	$125
5	$140
6	$155

15.

Charges for Overnight Pet Sitting

Number of Dogs (x)	Total Charges (y)
1	$65
2	$75
3	$85
4	$95
5	$105
6	$115

18.

Ernie's Electric Repair

Number of Hours (x)	Total Charges (y)
1	$120
2	$205
3	$290
4	$375
5	$460
6	$545

11.9 Function Mapping

Functional relationships can be represented in a variety of ways. One way is to draw a picture that shows how ordered pairs are formed. These are called **function maps**. Each element in the domain of a relation is shown on the left and each element of the range is shown on the right. The arrows create the ordered pairs by "mapping" each element of the domain onto the range. Remember, the relation is a function if each element in the domain is paired with exactly one element of the range.

In the examples below, the arrows from different elements in the domain can map to the same element in the range, but the same element in the domain **CANNOT** map to different elements in the range.

Correct Mapping   Incorrect Mapping

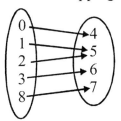 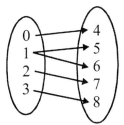

Remember, from the section on functions from ordered pairs, you could write out the ordered pairs from the maps and, if the same x-coordinate has more than one y-coordinate, then the relation is not a function. In the mapping above on the right, the ordered pairs mapped are $(0, 4)$, $(1, 5)$, $(1, 6)$, $(2, 7)$, and $(3, 8)$. Because the x-coordinate $\{1\}$ has two different y-coordinates $\{5, 6\}$, we see once again that the relation is not a function.

1. Which set of ordered pairs completely express the relation shown?

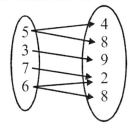

(A) $(5, 4)$, $(3, 9)$, $(7, 2)$, $(6, 8)$
(B) $(5, 8)$, $(3, 9)$, $(7, 2)$, $(6, 2)$
(C) $(4, 5)$, $(9, 3)$, $(2, 7)$, $(8, 6)$, $(8, 5)$
(D) $(5, 4)$, $(5, 8)$, $(3, 9)$, $(7, 2)$, $(6, 2)$, $(6, 8)$

2. The relation mapped in Question 1 above:
   (A) is a function because all of the x-coordinates are mapped to a y-coordinate
   (B) is a function because all of the y-coordinates have at least one x-coordinate
   (C) is not a function because all elements in the range are mapped
   (D) is not a function because at least one element in the domain is mapped to two different elements in the range

Chapter 11 Review

1. What is the domain of the following relation? $\{(-1,2),(2,5),(4,9),(6,11)\}$

2. What is the range of the following relation? $\{(0,-2),(-1,-4),(-2,6),(-3,-8)\}$

3. Find the range of the relation $y = 5x$ for the domain $\{0,1,2,3,4\}$.

4. Find the range of the relation $y = \dfrac{3(x-2)}{5}$ for the domain $\{-8,-3,7,12,17\}$.

5. Find the range of the relation $y = 10 - 2x$ for the domain $\{-8,-4,0,4,8\}$.

6. Find the range of the relation $y = \dfrac{4+x}{3}$ for the domain $\{-7,-1,2,5,8\}$.

For each of the following relations given in questions 7–11, write F if it is a function and NF if it is not a function.

7. $\{(1,2),(2,2),(3,2)\}$

8. $\{(-1,0),(0,1),(1,2),(2,3)\}$

9. $\{(2,1),(2,2),(2,3)\}$

10. $\{(1,7),(2,5),(3,6),(2,4)\}$

11. $\{(0,-1),(1,2),(-2,-3),(-3,-4)\}$

For questions 12–17, find the range of the following functions for the given value of the domain.

12. For $g(x) = 2x^2 - 4x$; find $g(-1)$

13. For $h(x) = 3x(x-4)$; find $h(3)$

14. For $f(n) = \dfrac{1}{n+3}$; find $f(4)$

15. For $G(n) = \dfrac{2-n}{2}$; find $G(8)$

16. For $H(x) = 2x(x-1)$; find $H(4)$

17. For $f(x) = 7x^2 + 3x - 2$; find $f(2)$

Graph each function.

18. $f(x) = -|2x|$

19. $f(x) = \frac{1}{2}x^2$

20. $f(x) = |-6x|$

21. $f(x) = \frac{1}{7}x$

22. $f(x) = -3\sqrt{x}$

23. $f(x) = -\frac{1}{8}\sqrt{x}$

Use the table to find the equation.

24.

x	-2	-1	0	1	2
y	4	5	6	7	8

25.

x	-2	-1	0	1	2
y	-4	-3.5	-3	-2.5	-2

26.

x	-2	-1	0	1	2
y	5	4	3	2	1

27.

x	-2	-1	0	1	2
y	15	16	17	18	19

Find the domain and range of each graph in questions 28 and 29.

28.

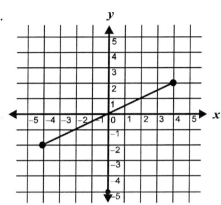

29.

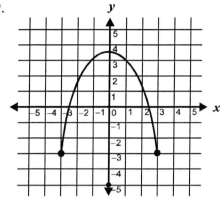

For questions 30 and 31, determine whether or not the graphs are functions.

30.

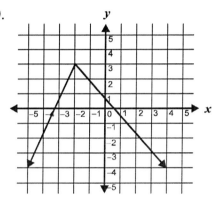

31.

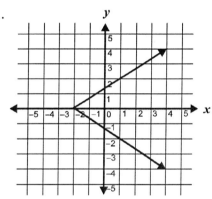

Fill in the following function tables.

32. Rule: $y = 3(x - 5)$

x	y
1	
2	
3	
4	
5	

34. Rule: $y = \dfrac{(6 + x)}{x}$

x	y
-2	
-1	
1	
2	
5	

36. Rule: $y = x(x + 1)$

x	y
1	
2	
3	
4	
5	

33. Rule: $\dfrac{(4 - 2n)}{2}$

n	$f(n)$
0	
1	
2	
3	
4	

35. Rule: $2n(n + 1)$

n	$f(n)$
0	
1	
2	
3	
4	

37. Rule: $6n - 3$

n	$f(n)$
2	
3	
4	
5	
6	

Chapter 11 Test

1. Which of the following relations is a function?

    A $\{(0, -1)(1, -1)(0, -2)\}$
    B $\{(-1, 1)(-1, -1)(0, 0)\}$
    C $\{(2, 1)(1, 0)(0, -1)\}$
    D $\{(-1, 1)(-1, 0)(-1, -1)\}$

2. Find the range of the following function for the domain $\{-2, -1, 0, 3\}$.

$$y = \frac{2 + x}{4}$$

    A $\left\{0, \dfrac{3}{4}, 1, \dfrac{5}{4}\right\}$

    B $\left\{0, \dfrac{1}{4}, \dfrac{1}{2}, \dfrac{5}{4}\right\}$

    C $\left\{1, -\dfrac{1}{4}, \dfrac{1}{2}, \dfrac{5}{4}\right\}$

    D $\left\{\dfrac{1}{4}, \dfrac{3}{4}, \dfrac{1}{2}, \dfrac{5}{4}\right\}$

3. Which of the following graphs is a function?

    A
    B
    C
    D

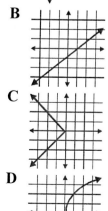

4. The following graph depicts the height of a projectile as a function of time.

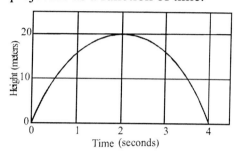

What is the domain (D) of this function?

    A 0 meters $\leq$ D $\leq$ 20 meters
    B 4 seconds $\leq$ D $\leq$ 20 meters
    C 20 meters $\leq$ D $\leq$ 4 seconds
    D 0 seconds $\leq$ D $\leq$ 4 seconds

5. What is the range of the following relation?

$\{(1, 2)(4, 9)(7, 8)(10, 13)\}$

    A $\{1, 4, 7, 10\}$
    B $\{2, 9, 8, 13\}$
    C $\{3, 13, 15, 23\}$
    D $\{1, 3, 1, 3\}$

6. What is the domain of the function graphed below?

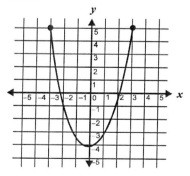

    A -3
    B 3
    C $-4 \leq x \leq 5$
    D $-3 \leq x \leq 3$

7. Which ordered pair could be part of this function: $(2, 4)$, $(3, 7)$, $(9, 1)$?

   A $(4, 1)$

   B $(3, 11)$

   C $(2, 9)$

   D $(9, 0)$

8. What is the range of the function graphed below?

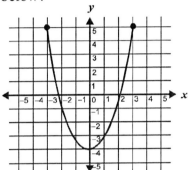

   A -3

   B 3

   C $-4 \leq y \leq 5$

   D None of the above

9. What is the domain of the relation $\{(2, 4)\, (3, 7)\, (4, 9)\, (6, 11)\}$?

   A $\{2, 3, 4, 6\}$

   B $\{4, 7, 9, 11\}$

   C $(2, 4)$

   D $(2, 4)\, (3, 7)$

10. Find the range of $y = 5x$ with a domain $\{1, 2, 3, 4\}$.

    A $1, 2, 3, 4$

    B $5, 6, 7, 8$

    C $5, 10, 15, 20$

    D $5, 15, 25, 35$

11. Which is not a function?

    A

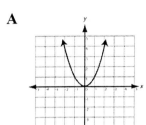

    B

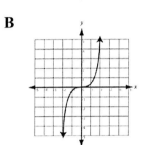

    C

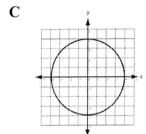

    D

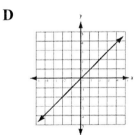

12. If $f(x) = 2x + 1$, what is $f(-3)$?

    A 5

    B -7

    C -5

    D 6

13. Which equation is represented by the function table below?

x	$f(x)$
1	8
2	14
3	20
4	26

A $f(x) = 3x - 1$

B $f(x) = \frac{1}{2}x + 1$

C $f(x) = 7x - \frac{3}{7}$

D $f(x) = 6x + 2$

14. Which of the following mapping represents the set of ordered pairs $\{(1,9),(2,2),(3,1),(4,2)\}$?

A

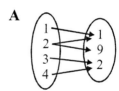

B

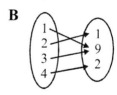

C

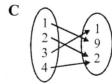

D

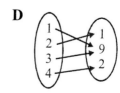

15. Which is the graph of $f(x) = x^2$?

A

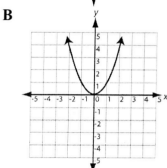

B

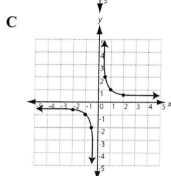

C

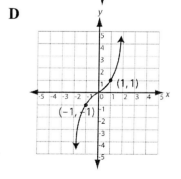

D

Chapter 12
Statistics

This chapter covers the following Alabama objectives and standards in mathematics:

HSGE
Mathematics

	Objective(s)
Standard VII	5

12.1   Range

In **statistics**, the difference between the largest number and the smallest number in a list is called the **range**.

Example 1:   Find the range of the following list of numbers: 16, 73, 26, 15, and 35.

The largest number is 73, and the smallest number is 15. $73 - 15 = 58$
The range is 58.

Find the range for each list of numbers below.

1.		2.		3.		4.		5.		6.		7.	
	21		6		89		41		23		2		77
	51		7		22		3		20		38		94
	48		31		65		56		64		29		27
	42		55		36		41		38		33		46
	12		8		20		19		21		59		63

8.		9.		10.		11.		12.		13.		14.	
	51		65		84		84		21		45		62
	62		54		59		65		78		57		39
	32		56		48		32		6		57		96
	16		5		21		50		97		14		45
	59		63		80		71		45		61		14

15. 2, 15, 3, 25, and 17

16. 15, 48, 52, 41, and 8

17. 54, 74, 2, 86, and 75

18. 15, 61, 11, 22, and 65

19. 33, 18, 65, 12, and 74

20. 47, 12, 33, 25, and 19

21. 56, 10, 33, 7, 16, and 5

22. 46, 25, 78, 49, and 6

23. 45, 75, 63, and 21

24. 97, 23, 56, 12, and 66

25. 87, 44, 63, and 12

26. 84, 55, 66, 38, and 31

27. 35, 44, 81, 99, and 78

28. 95, 54, 62, 14, 8, and 3

12.2   Mean

In statistics, the mean is the same as the average. To find the mean of a list of numbers, first add together all of the numbers in the list, and then divide by the number of items in the list.

Example 2:   Find the mean of 38, 72, 110, 548.

Step 1:   First add: $38 + 72 + 110 + 548 = 768$

Step 2:   There are 4 numbers in the list so divide the total by 4. $768 \div 4 = 192$
The mean is 192.

Practice finding the mean (average). Round to the nearest tenth if necessary.

1. Dinners served:
   489 561 522 450

2. Prices paid for shirts:
   $4.89 $9.97 $5.90 $8.64

3. Piglets born:
   23 19 15 21 22

4. Student absences:
   6 5 13 8 9 12 7

5. Paychecks:
   $89.56 $99.99 $56.54

6. Choir attendance:
   56 45 97 66 70

7. Long distance calls:
   33 14 24 21 19

8. Train boxcars:
   56 55 48 61 51

9. Cookies eaten:
   5 6 8 9 2 4 3

Find the mean (average) of the following word problems.

10. Val's science grades are 95, 87, 65, 94, 78, and 97. What is her average?

11. Ann runs a business from her home. The number of orders for the last 7 business days are 17, 24, 13, 8, 11, 15, and 9. What is the average number of orders per day?

12. Melissa tracks the number of phone calls she has per day: 8, 2, 5, 4, 7, 3, 6, 1. What is the average number of calls she receives?

13. The Cheese Shop tracks the number of lunches they served this week: 42, 55, 36, 41, 38, 33, and 46. What is the average number of lunches served?

14. Leah drives 364 miles in 7 hours. What is her average miles per hour?

15. Tim saves $680 in 8 months. How much does his savings average each month?

16. Ken makes 117 passes in 13 games. How many passes does he average per game?

12.3   Finding Data Missing from the Mean

Example 3:   Mara knew she had an 88 average in her biology class, but she lost one of her papers. The three papers she could find had scores of 98%, 84%, and 90%. What was the score on her fourth paper?

Step 1:   Figure the total score on four papers with an 88% average. $0.88 \times 4 = 3.52$

Step 2:   Add together the scores from the three papers you have. $0.98 + 0.84 + 0.9 = 2.72$

Step 3:   Subtract the scores you know from the total score. $3.52 - 2.72 = 0.80$. She had 80% on her fourth paper.

Find the data missing from the following problems.

1. Gabriel earns 87% on his first geography test. He wants to keep a 92% average. What does he need to get on his next test to bring his average up?

2. Rian earned $68.00 on Monday. How much money must she earn on Tuesday to have an average of $80 earned for the two days?

3. Haley, Chuck, Dana, and Chris enter a contest to see who could bake the most chocolate chip cookies in an hour. They bake an average of 75 cookies. Haley bakes 55, Chuck bakes 70, and Dana bakes 90. How many does Chris bake?

4. Four wrestlers make a pact to lose some weight before the competition. They lose an average of 7 pounds each over the course of 3 weeks. Carlos loses 6 pounds, Steve loses 5 pounds, and Greg loses 9 pounds. How many pounds does Wes lose?

5. Three boxes are ready for shipment. The boxes average 26 pounds each. The first box weighs 30 pounds; the second box weighs 25 pounds. How much does the third box weigh?

6. The five jockeys running in the next race average 92 pounds each. Nicole weighs 89 pounds. Jon weighs 95 pounds. Jenny and Kasey weigh 90 pounds each. How much does Jordan weigh?

7. Jessica makes three loaves of bread that weigh a total of 45 ounces. What is the average weight of each loaf?

8. Celeste makes scented candles to give away to friends. She has 2 pounds of candle wax which she melted, scented, and poured into 8 molds. What is the average weight of each candle?

9. Each basketball player has to average a minimum of 5 points a game for the next three games to stay on the team. Ben is feeling the pressure. He scored 3 points the first game and 2 points the second game. How many points does he need to score in the third game to stay on the team?

12.4   Median

In a list of numbers ordered from lowest to highest, the **median** is the middle number. To find the **median**, first arrange the numbers in numerical order. If there is an odd number of items in the list, the **median** is the middle number. If there is an even number of items in the list, the **median** is the **average of the two middle numbers.**

Example 4:   Find the median of 42, 35, 45, 37, and 41.

Step 1:   Arrange the numbers in numerical order: 35 37 $\boxed{41}$ 42 45

Step 2:   Find the middle number. The median is 41.

Example 5:   Find the median of 14, 53, 42, 6, 14, and 46.

Step 1:   Arrange the numbers in numerical order: 6 14 $\boxed{14\ 42}$ 46 53.

Step 2:   Find the average of the two middle numbers.
$(14 + 42) \div 2 = 28$. The median is 28.

Circle the median in each list of numbers.

1. 35, 55, 40, 30, and 45

2. 7, 2, 3, 6, 5, 1, and 8

3. 65, 42, 60, 46, and 90

4. 15, 16, 19, 25, 20

5. 75, 98, 87, 65, 82, 88, 100

6. 33, 42, 50, 22, and 19

7. 401, 758, and 254

8. 41, 23, 14, 21, and 19

9. 5, 8, 10, 13, 1, and 8

10.	11.	12.	13.	14.	15.	16.
19	9	45	52	20	8	15
14	3	32	54	21	17	40
12	10	66	19	25	13	42
15	17	55	63	18	14	32
18	6	61	20	16	22	28

Find the median in each list of numbers.

17. 10, 8, 21, 14, 9, and 12

18. 43, 36, 20, and 40

19. 5, 24, 9, 18, 12, and 3

20. 48, 13, 54, 82, 90, and 7

21. 23, 21, 36, and 27

22. 9, 4, 3, 1, 6, 2, 10, and 12

23.	24.	25.	26.	27.	28.
2	11	13	75	48	22
10	22	15	62	45	19
6	25	9	60	52	15
18	28	35	52	30	43
20	10	29	80	35	34
23	23	33	50	58	28

12.5   Mode

In statistics, the mode is the number that occurs most frequently in a list of numbers.

Example 6:   Exam grades for a math class were as follows:
70 88 92 85 99 85 70 85 99 100 88 70 99 88 88 99 88 92 85 88

Step 1:   Count the number of times each number occurs in the list.

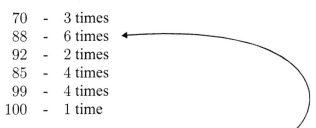

70 - 3 times
88 - 6 times
92 - 2 times
85 - 4 times
99 - 4 times
100 - 1 time

Step 2:   Find the number that occurs most often.
The mode is 88 because it is listed 6 times. No other number is listed as often.

Find the mode in each of the following lists of numbers.

1.	2.	3.	4.	5.	6.	7.
88	54	21	56	64	5	12
15	42	16	67	22	4	41
88	44	15	67	22	9	45
17	56	78	19	15	8	32
18	44	21	56	14	4	16
88	44	16	67	14	7	12
17	56	21	20	22	4	12

8. 48, 32, 56, 32, 56, 48, 56

9. 12, 16, 54, 78, 16, 25, 20

10. 5, 4, 8, 3, 4, 2, 7, 8, 4, 2

11. 11, 9, 7, 11, 7, 5, 7, 7, 5

12. 84, 22, 79, 22, 87, 22, 22

13. 95, 87, 65, 94, 78, 95

14. 8, 2, 5, 4, 7, 2, 3, 6, 1

15. 89, 7, 11, 89, 17, 56

16. 15, 48, 52, 41, 8, 48

17. 22, 45, 48, 12, 22, 41, 22

18. 62, 44, 78, 62, 54, 44, 62

19. 54, 22, 54, 78, 22, 78, 22

20. 14, 17, 33, 21, 33, 17, 33

21. 65, 51, 8, 21, 8, 65, 70, 8

22. 17, 24, 13, 8, 11, 8, 15, 9

23. 51, 45, 84, 51, 65, 74, 51

24. 8, 74, 65, 15, 9, 10, 74

25. 62, 54, 2, 7, 89, 2, 7, 54, 2

12.6   Applying Measures of Central Tendency

On the AL High School Graduation exam, you may be asked to solve real-world problems involving measures of central tendency.

Example 7:   Aida is shopping around for the best price on a 17" computer monitor. She travels to seven stores and finds the following prices: $199, $159, $249, $329, $199, $209, and $189. When Aida goes to the eighth and final store, she finds the price for the 17" monitor is $549. Which of the measures of central tendency, mean, median, or mode, changes the most as a result of the last price Aida finds?

Step 1:   **Solve for all three measures of the seven values.**

Mean: $\dfrac{\$199 + \$159 + \$249 + \$329 + \$199 + \$209 + \$189}{7} = \$219$

Median: From least to greatest: $159, $189, $199, $199, $209, $249, $329. The 4th value $= \$199$

Mode: The number repeated the most is $199.

Step 2:   **Find the mean, median, and mode with the eighth value included.**

Mean: $\dfrac{\$199 + \$159 + \$249 + \$329 + \$199 + \$209 + \$189 + \$549}{8} = \$260.25$

Median: $159, $189, $199, $199, $209, $249, $329, $549. The avg. of 4th and 5th number $= \$204$

Mode: The number still repeated most is $199.

Answer: The measure which changed the most by adding the 8th value is the **mean**.

1. The Realty Company has the selling prices for 10 houses sold during the month of July. The following prices are given in thousands of dollars:

   176   89   525   125   107   100   525   61   75   114

   Find the mean, median, and mode of the selling prices. Which measure is most representative for the selling price of such homes? Explain.

2. A soap manufacturing company wants to know if the weight of its product is on target, meaning 4.75 oz. With that purpose in mind, a quality control technician selects 15 bars of soap from production, 5 from each shift, and finds the following weights in oz.

   1st shift: 4.76, 4.75, 4.77, 4.77, 4.74
   2nd shift: 4.72, 4.72, 4.75, 4.76, 4.73
   3rd shift: 4.76, 4.76, 4.77, 4.76, 4.76

   (A) What are the values for the measures of central tendency for the sample from each shift?
   (B) Find the mean, median, and mode for the 24 hour production sample.
   (C) Which measure is the most accurate measure of central tendency for the 24 hour production?
   (D) Find the range of values for each shift. Is the range an effective tool for drawing a conclusion in this case? Why or why not?

12.7 Tally Charts and Frequency Tables

Large lists can be tallied in a chart. To make a **tally chart**, record a tally mark in a chart for each time a number is repeated. To make a **frequency table**, count the times each number occurs in the list, and record the frequency.

Example 8: The age of each student in grades 6–8 in a local middle school are listed below. Make a tally chart and a frequency table for each age.

Student Ages						
grades 6–8						
10	11	11	12	14	12	11
13	13	13	12	14	11	12
12	14	12	10	15	11	13
12	10	12	11	12	13	12
13	12	13	12	11	10	13
14	14	11	15	12	13	14
12	11	14	12	11	13	

TALLY CHART	
Age	Tally
10	IIII
11	HHH HHH
12	HHH HHH HHH
13	HHH HHH
14	HHH II
15	II

FREQUENCY TABLE	
Age	Frequency
10	4
11	10
12	15
13	10
14	7
15	2

Make a chart to record tallies and frequencies for the following problems.

1. The sheriff's office monitors the speed of cars traveling on Turner Road for one week. The following data is the speed of each car that travels Turner Road during the week. Tally the data in 10 miles per hour (mph) increments starting with 10–19 mph, and record the frequency in a chart.

Car Speed, mph									
45	52	47	35	48	50	51	43	52	41
40	51	32	24	55	41	32	33	45	
36	39	49	52	34	28	39	47	56	
29	15	63	42	35	42	58	59	35	
39	41	25	34	22	16	40	31	55	
55	10	46	38	50	52	48	36	65	
21	32	36	41	52	49	45	32	20	

Speed	Tally	Frequency
10-19		
20-29		
30-39		
40-49		
50-59		
60-69		

2. The following data gives final math averages for Ms. Kirby's class. In her class, an average of 90–100 is an A, 80–89 is B, 70–79 is a C, 60–69 is a D, and an average below 60 is an F. Tally and record the frequency of A's, B's, C's D's, and F's.

Final Math Averages									
85	92	87	62	75	84	96	52	31	79
45	77	98	75	71	79	85	82	86	76
87	74	76	68	93	77	65	84	89	
79	65	77	82	86	84	92	60	65	
99	75	88	74	79	80	63	84	69	
87	90	75	81	73	69	73	75	75	

Grade	Tally	Frequency
A		
B		
C		
D		
F		

Chapter 12 Review

Find the mean, median, mode, and range for each of the following sets of data. Fill in the table below.

❶ Miles Run by Track Team Members	
Jeff	24
Eric	20
Craig	19
Simon	20
Elijah	25
Rich	19
Marcus	20

❷ 1992 SUMMER OLYMPIC GAMES
Gold Medals Won

Unified Team	45	Hungary	11
United States	37	South Korea	12
Germany	33	France	8
China	16	Australia	7
Cuba	14	Japan	3
Spain	13		

❸ Hardware Store Payroll June Week 2	
Erica	$280
Dane	$206
Sam	$240
Nancy	$404
Elsie	$210
Gail	$305
David	$280

Data Set Number	Mean	Median	Mode	Range
❶				
❷				
❸				

4. Jenica bowls three games and scores an average of 116 points per game. She scores 105 on her first game and 128 on her second game. What does she score on her third game?

5. Concession stand sales for each game in season are $320, $540, $230, $450, $280, and $580. What is the mean sales per game?

6. Cendrick D'Amitrano works Friday and Saturday delivering pizza. He delivers 8 pizzas on Friday. How many pizzas must he deliver on Saturday to average 11 pizzas per day?

7. Long cooks three Vietnamese dinners that weigh a total of 40 ounces. What is the average weight for each dinner?

8. The Swamp Foxes scored an average of 7 points per soccer game. They scored 9 points in the first game, 4 points in the second game, and 5 points in the third game. What was their score for their fourth game?

9. Shondra is 66 inches tall, and DeWayne is 74 inches tall. How tall is Michael if the average height of these three students is 73 inches?

10. The 6th grade did a survey on the number of pets each student had at home. The following chart displays the data produced by the survey. Create a frequency table for this data.

NUMBER OF PETS PER STUDENT																								
0	2	6	2	1	0	4	2	3	3	0	2	3	5	1	4	2	0	5	2	3	3	4	3	6	2
5	1	2	3	5	6	3	2	2	5	2	3	4	3	0	1	4	1	2	4	5	7	6	1	4	7

Chapter 12 Test

1. What is the mean of 36, 54, 66, 45, 36, 36, and 63?

   A 36
   B 45
   C 48
   D 63

2. The student council surveyed the student body on favorite lunch items. The frequency chart below shows the results of the survey.

Favorite Lunch Item	Frequency
corndog	140
hamburger	245
hotdog	210
pizza	235
spaghetti	90
other	65

   Which lunch item indicates the mode of the data?

   A other
   B hotdog
   C hamburger
   D corndog

3. Concession stand sales for the first 6 games of the season averaged $400.00. If the total sales of the first 5 games were $320, $540, $230, $450, and $280, what were the total sales for the sixth game?

   A $230
   B $350
   C $364
   D $580

4. What is the mean of 12, 23, 8, 26, 37, 11, and 9?

   A 12
   B 29
   C 18
   D 19

5. Which of the following sets of numbers has a median of 42?

   A $\{60, 42, 37, 22, 19\}$
   B $\{16, 28, 42, 48\}$
   C $\{42, 64, 20\}$
   D $\{12, 42, 40, 50\}$

6. A neighborhood surveyed the times of day people water their lawns and tallied the data below.

Time	Tally
midnight - 3:59 a.m.	||
4:00 a.m. - 7:59 a.m.	JHT I
8:00 a.m. - 11:59 a.m.	JHT IIII
noon - 3:59 p.m.	JHT
4:00 p.m. - 7:59 p.m.	JHT JHT
8:00 p.m. - 11:59 p.m.	JHT III

   If you wanted to find which was the most popular time of day to water the lawn, it would be best to find the _____ of data.

   A mean
   B median
   C range
   D mode

7. Examine the following two data sets:

   Set #1: 49, 55, 68, 72, 98

   Set #2: 20, 36, 47, 68, 75, 82, 89

   Which of the following statements is true?

   A They have the same mode.
   B They have the same median.
   C They have the same mean.
   D None of the above.

8. Which of the following sets of numbers has a range of 51?

   A {29, 19, 72, 68, 39}
   B {81, 85, 37, 41, 60}
   C {17, 12, 9, 47, 82}
   D {62, 86, 44, 78, 95}

Use the figure for questions 9–10.

Bob	7	4	6	8	7	5	3
Ann	6	5	6	7	9	5	4

9. Find the means. Who produced more?

   A Ann, 6
   B Ann, 5.7
   C Bob, 5.7
   D Bob, 6

10. What is the mode for Bob's numbers?

   A 4
   B 5
   C 6
   D 7

Chapter 13
Probability

HSGE

Mathematics

This chapter covers the following Alabama objectives and standards in mathematics:

	Objective(s)
Standard VII	6

13.1   Probability

Probability is the chance something will happen. Probability is most often expressed as a fraction, a decimal, a percent, or can also be written out in words

Example 1:   Billy has 3 red marbles, 5 white marbles, and 4 blue marbles on the floor. His cat comes along and bats one marble under the chair. What is the **probability** it is a red marble?

Step 1:   The number of red marbles will be the top number of the fraction. $\longrightarrow \dfrac{\mathbf{3}}{\mathbf{12}}$
Step 2:   The total number of marbles is the bottom number of the fraction.
The answer may be expressed in lowest terms. $\frac{3}{12} = \frac{1}{4}$.
Expressed as a decimal, $\frac{1}{4} = 0.25$, as a percent, $\frac{1}{4} = 25\%$, and written out in words, $\frac{1}{4}$ is one out of four.

Example 2:   Determine the probability that the pointer will stop on a shaded wedge or the number 1.

Step 1:   Count the number of possible wedges that the spinner can stop on to satisfy the above problem. There are 5 wedges that satisfy it (4 shaded wedges and one number 1). The top number of the fraction is 5.

Step 2:   Count the total number of wedges, 7. The bottom number of the fraction is 7. The answer is $\frac{5}{7}$ or **five out of seven.**

Example 3:   Refer to the spinner above. If the pointer stops on the number 7, what is the probability that it will **not** stop on 7 the next time?

Step 1:   Ignore the information that the pointer stopped on 7 the previous spin. The probability of the next spin does not depend on the outcome of the previous spin. Simply find the probability that the spinner will **not** stop on 7. If P is the probability of an event occurring, $1 - P$ is the probability of an event **not** occurring. In this example, the probability of the spinner landing on 7 is $\frac{1}{7}$.

Step 2:   The probability that the spinner will not stop on 7 is $1 - \frac{1}{7}$ which equals $\frac{6}{7}$. The answer is $\frac{6}{7}$ or **six out of seven**.

Find the probability of the following problems. Express the answer as a percent.

1. A computer chooses a random number between 1 and 50. What is the probability that you will guess the same number that the computer chose in 1 try?

2. There are 24 candy-coated chocolate pieces in a bag. Eight have defects in the coating that can be seen only with close inspection. What is the probability of pulling out a defective piece without looking?

3. Seven sisters have to choose which day each will wash the dishes. They put equal-sized pieces of paper in a hat, each labeled with a day of the week. What is the probability that the first sister who draws will choose a weekend day?

4. For his garden, Clay has a mixture of 12 white corn seeds, 24 yellow corn seeds, and 16 bicolor corn seeds. If he reaches for a seed without looking, what is the probability that Clay will plant a bicolor corn seed first?

5. Mom just got a new department store credit card in the mail. What is the probability that the last digit is an odd number?

6. Alex has a paper bag of cookies that holds 8 chocolate chip, 4 peanut butter, 6 butterscotch chip, and 12 ginger. Without looking, his friend John reaches in the bag for a cookie. What is the probability that the cookie is peanut butter?

7. An umpire at a little league baseball game has 14 balls in his pockets. Five are brand A, 6 are brand B, and 3 are brand C. What is the probability that the next ball he throws to the pitcher is a brand C ball?

8. What is the probability that the spinner's arrow will land on an even number?

9. The spinner in the problem above stopped on a shaded wedge on the first spin and stopped on the number 2 on the second spin. What is the probability that it will not stop on a shaded wedge or on the 2 on the third spin?

10. A company is offering 1 grand prize, 3 second place prizes, and 25 third place prizes based on a random drawing of contest entries. If your entry is one of the 500 total entries, what is the probability you will win a third place prize?

11. In the contest problem above, what is the probability that you will win the grand prize or a second place prize?

12. A box of a dozen doughnuts has 3 lemon cream-filled, 5 chocolate cream-filled, and 4 vanilla cream-filled. If the doughnuts look identical, what is the probability of picking a lemon cream-filled?

13.2   Independent and Dependent Events

In mathematics, the outcome of an event may or may not influence the outcome of a second event. If the outcome of one event does not influence the outcome of the second event, these events are **independent**. However, if one event has an influence on the second event, the events are **dependent**. When someone needs to determine the probability of two events occurring, he or she will need to use an equation. These equations will change depending on whether the events are independent or dependent in relation to each other. When finding the probability of two **independent** events, multiply the probability of each favorable outcome together.

Example 4:   One bag of marbles contains 1 white, 1 yellow, 2 blue, and 3 orange marbles. A second bag of marbles contains 2 white, 3 yellow, 1 blue, and 2 orange marbles. What is the probability of drawing a blue marble from each bag?

Solution: Probability of favorable outcomes

Bag 1: $\dfrac{2}{7}$

Bag 2: $\dfrac{1}{8}$

Probability of a blue marble from each bag: $\dfrac{2}{7} \times \dfrac{1}{8} = \dfrac{2}{56} = \dfrac{1}{28}$

In order to find the probability of two **dependent** events, you will need to use a different set of rules. For the first event, you must divide the number of favorable outcomes by the number of possible outcomes. For the second event, you must subtract one from the number of favorable outcomes **only if** the favorable outcome is the **same**. However, you must subtract one from the number of total possible outcomes. Finally, you must multiply the probability for event one by the probability for event two.

Example 5:   One bag of marbles contains 3 red, 4 green, 7 black, and 2 yellow marbles. What is the probability of drawing a green marble, removing it from the bag, and then drawing another green marble?

	Favorable Outcomes	Total Possible Outcomes
Draw 1	4	16
Draw 2	3	15
Draw 1 × Draw 2	12	240

Answer: $\dfrac{12}{240}$ or $\dfrac{1}{20}$

Example 6:   Using the same bag of marbles, what is the probability of drawing a red marble, keeping it and then drawing a black marble?

	Favorable Outcomes	Total Possible Outcomes
Draw 1	3	16
Draw 2	7	15
Draw 1 × Draw 2	21	240

Answer $\dfrac{21}{240}$ or $\dfrac{7}{80}$

Find the probability of the following problems. Express the answer as a fraction.

1. Prithi has two boxes. Box 1 contains 3 red, 2 silver, 4 gold, and 2 blue combs. She also has a second box containing 1 black and 1 clear brush. What is the probability that Prithi selects a red comb from box 1 and a black brush from box 2?

2. Steve Marduke has two spinners in front of him. The first one is numbered 1–6, and the second is numbered 1–3. If Steve spins each spinner once, what is the probability that the first spinner will show an odd number and the second spinner will show a "1"?

3. Carrie McCallister flips a coin twice and gets heads both times. What is the probability that Carrie will get tails the third time she flips the coin?

4. Artie Drake turns a spinner which is evenly divided into 11 sections numbered 1–11. On the first spin, Artie's pointer lands on "8". What is the probability that the spinner lands on an even number the second time he turns the spinner?

5. Leanne Davis plays a game with a street entertainer. In this game, a ball is placed under one of three coconut halves. The vendor shifts the coconut halves so quickly that Leanne can no longer tell which coconut half contains the ball. She selects one and misses. The entertainer then shifts all three around once more and asks Leanne to pick again. What is the probability that Leanne will select the coconut half containing the ball?

6. What is the probability that Jane Robelot reaches into a bag containing 1 daffodil and 2 gladiola bulbs and pulls out a daffodil bulb, and then reaches into a second bag containing 6 tulip, 3 lily, and 2 gladiola bulbs and pulls out a lily bulb?

7. Terrell casts his line into a pond containing 7 catfish, 8 bream, 3 trout, and 6 northern pike. He immediately catches a bream. What are the chances that Terrell will catch a second bream the next time he casts his line?

8. Gloria Quintero enters a contest in which the person who draws his or her initials out of a box containing all 26 letters of the alphabet wins the grand prize. Gloria reaches in, draws a "G", keeps it, then draws another letter. What is the probability that Gloria will next draw a "Q"?

9. Vince Macaluso is pulling two socks out of a washing machine in the dark. The washing machine contains three tan, one white, and two black socks. If Vince reaches in and pulls out the socks one at a time, what is the probability that he will pull out two tan socks on his first two tries?

10. John Salome has a bag containing 2 yellow plums, 2 red plums, and 3 purple plums. What is the probability that he reaches in without looking and pulls out a yellow plum and eats it, then reaches in again without looking and pulls out a red plum to eat?

13.3 More Probability

Example 7: You have a cube with one number, 1, 2, 3, 4, 5 and 6 painted on each face of the cube. What is the probability that if you throw the cube 3 times, you will get the number 2 each time?

If you roll the cube once, you have a 1 in 6 chance of getting the number 2. If you roll the cube a second time, you again have a 1 in 6 chance of getting the number 2. If you roll the cube a third time, you again have a 1 in 6 chance of getting the number 2. The probability of rolling the number 2 three times in a row is:

$$\frac{1}{6} \times \frac{1}{6} \times \frac{1}{6} = \frac{1}{216}$$

Find the probability that each of the following events will occur.

There are 10 balls in a box, each with a different digit on it: 0, 1, 2, 3, 4, 5, 6, 7, 8, & 9. A ball is chosen at random and then put back in the box.

1. What is the probability that if you pick out a ball 3 times, you will get number 7 each time?

2. What is the probability you will pick a ball with 5, then 9, and then 3?

3. What is the probability that if you pick out a ball 4 times, you will always get an odd number?

4. A couple has 4 children ages 9, 6, 4, and 1. What is the probability that they are all girls?

There are 26 letters in the alphabet, allowing a different letter to be on each of 26 cards. The cards are shuffled. After each card is chosen at random, it is put back in the stack of cards, and the cards are shuffled again.

5. What is the probability that when you pick 3 cards, you would draw first a "y", then and "e", and then an "s"?

6. What is the probability that you would draw 4 cards and get the letter "z" each time?

7. What is the probability that you draw twice and get a letter in the word "random" both times?

8. If you flip a coin 3 times, what is the probability you will get heads every time?

9. Marie is clueless about 4 of her multiple-choice answers. The possible answers are A, B, C, D, E, or F. What is the probability that she will guess all four answers correctly?

Chapter 13 Review

1. There are 50 students in the school orchestra in the following sections:

   25 string section
   15 woodwind
    5 percussion
    5 brass

   One student will be chosen at random to present the orchestra director with an award. What is the probability the student will be from the woodwind section?

2. Fluffy's cat treat box contains 6 chicken-flavored treats, 5 beef-flavored treats, and 7 fish-flavored treats. If Fluffy's owner reaches in the box without looking and chooses one treat, what is the probability that Fluffy will get a chicken-flavored treat?

3. The spinner in Figure A stopped on the number 5 on the first spin. What is the probability that it will not stop on 5 on the second spin?

Fig. A   Fig. B

4. Sherri turns the spinner in Figure B above 3 times. What is the probability that the pointer always lands on a shaded number?

5. Three cakes are sliced into 20 pieces each. Each cake contains 1 gold ring. What is the probability that one person who eats one piece of cake from each of the 3 cakes will find 3 gold rings?

6. Brianna tosses a coin 4 times. What is the probability she gets all tails?

Read the following, and answer questions 7–12.

There are 9 slips of paper in a hat, each with a number from 1 to 9. The numbers correspond to a group of students who must answer a question when the number for their group is drawn. Each time a number is drawn, the number is put back in the hat.

7. What is the probability that the number 6 will be drawn twice in a row?

8. What is the probability that the first 5 numbers drawn will be odd numbers?

9. What is the probability that the second, third, and fourth numbers drawn will be even numbers?

10. What is the probability that the first five times a number is drawn it will be the number 5?

11. What is the probability that the first five numbers drawn will be 1, 2, 3, 4, 5 in that order?

12. What are the odds against even group numbers 2, 4, 6, and 8 will be picked at any given drawing?

Solve the following word problems. For questions 13–15, write whether the problem is "dependent" or "independent."

13. Felix Perez reaches into a 10-piece puzzle and pulls out one piece at random. This piece has two places where it could connect to other pieces. What is the probability that he will select another piece which fits the first one if he selects the next piece at random?

14. Barbara Stein is desperate for a piece of chocolate candy. She reaches into a bag which contains 8 peppermint, 5 butterscotch, 7 toffee, 3 mint, and 6 chocolate pieces and pulls out a toffee piece. Disappointed, she throws it back into the bag and then reaches back in and pulls out one piece of candy. What is the probability that Barbara pulls out a chocolate piece on the second try?

15. Christen Solis goes to a pet shop and immediately decides to purchase a guppy she saw swimming in an aquarium. She reaches into the tank containing 5 goldfish, 6 guppies, 4 miniature catfish, and 3 minnows and accidently pulls up a goldfish. Breathing a sigh, Christen places the goldfish back in the water. The fish are swimming so fast, it is impossible to tell what fish Christen would catch. What is the probability that Christen will catch a guppy on her second try?

Chapter 13 Test

1. There are 10 boys and 12 girls in a class. If one student is selected at random from the class, what is the probability it is a girl?

   A $\frac{1}{2}$

   B $\frac{1}{22}$

   C $\frac{6}{11}$

   D $\frac{6}{5}$

2. David just got a new credit card in the mail. What is the probability the second digit of the credit card number is a 3?

   A $\frac{1}{3}$

   B $\frac{1}{10}$

   C $\frac{2}{3}$

   D $\frac{1}{5}$

3. Brenda has 18 fish in an aquarium. The fish are the following colors: 5 orange, 7 blue, 2 black, and 4 green. Brenda also has a trouble-making cat that has grabbed a fish. What is the probability the cat grabbed a green fish if all the fish are equally capable of avoiding the cat?

   A $\frac{2}{9}$

   B $\frac{1}{18}$

   C $\frac{2}{7}$

   D $\frac{1}{4}$

4. In problem number 3, what is the probability the cat will **not** grab an orange fish?

   A $\frac{1}{3}$

   B $\frac{13}{18}$

   C $\frac{3}{4}$

   D $\frac{13}{5}$

5. You have a cube with each face numbered 1, 2, 3, 4, 5, or 6. What is the probability if you roll the cube 4 times, you will get the number 5 each time?

   A $\frac{4}{1296}$

   B $\frac{4}{5}$

   C $\frac{1}{256}$

   D $\frac{1}{1296}$

6. Katie spun a spinner 15 times and recorded her results in a table below. The spinner was divided into 6 sections numbered 1–6. The results of the spins are shown below.

1	3	6	5	1
6	2	4	3	4
2	5	1	4	5

   Based on the results, how many times would 4 be expected to appear in 45 spins?

   A 9

   B 12

   C 15

   D 21

Chapter 14
Angles

This chapter covers the following Alabama objectives and standards in mathematics:

	Objective(s)
Standard VII	1

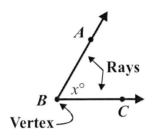

Angles are made up of two rays with a common endpoint. Rays are named by the endpoint B and another point on the ray. Ray $\overrightarrow{BA}$ and ray $\overrightarrow{BC}$ share a common endpoint.

Angles are usually named by three capital letters. The middle letter names the vertex. The angle to the left can be named $\angle ABC$ or $\angle CBA$. An angle can also be named by a lower case letter between the sides, $\angle x$, or by the vertex alone, $\angle B$.

A protractor, is used to measure angles. The protractor is divided evenly into a half circle of 180 degrees (180°). When the middle of the bottom of the protractor is placed on the vertex, and one of the rays of the angle is lined up with 0°, the other ray of the angle crosses the protractor at the measure of the angle. The angle below has the ray pointing left lined up with 0° (the outside numbers), and the other ray of the angle crosses the protractor at 55°. The angle measures 55°.

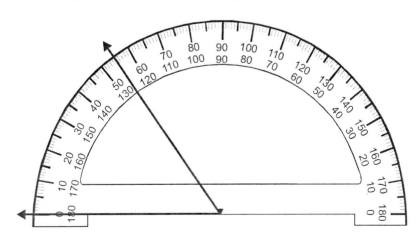

14.1   Types of Angles

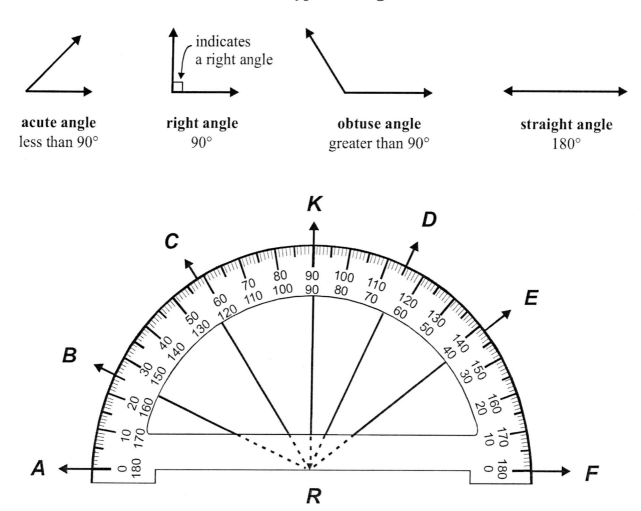

acute angle
less than 90°

right angle
90°

indicates
a right angle

obtuse angle
greater than 90°

straight angle
180°

Using the protractor above, find the measure of the following angles. Then, tell what type of angle it is: acute, right , obtuse, or straight.

		Measure	Type of Angle
1.	What is the measure of angle *ARF*?		
2.	What is the measure of angle *CRF*?		
3.	What is the measure of angle *BRF*?		
4.	What is the measure of angle *ERF*?		
5.	What is the measure of angle *ARB*?		
6.	What is the measure of angle *KRA*?		
7.	What is the measure of angle *CRA*?		
8.	What is the measure of angle *DRF*?		
9.	What is the measure of angle *ARD*?		
10.	What is the measure of angle *FRK*?		

14.2 Adjacent Angles

Adjacent angles are two angles that have the same vertex and share one ray. They do not share space inside the angles.

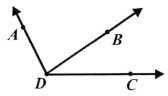

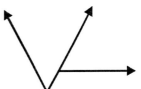

∠*ADB* is **adjacent to**∠*BDC*. However, ∠*ADB* is **not adjacent** to ∠*ADC* because adjacent angles do not share any space inside the angle

These two angles are **not adjacent**. They share a common ray but do not share the same vertex.

For each diagram below, name the angle that is adjacent to it.

1.

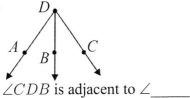

∠*CDB* is adjacent to ∠_____

5.

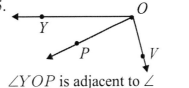

∠*YOP* is adjacent to ∠_____

2.

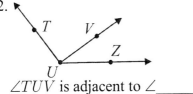

∠*TUV* is adjacent to ∠_____

6.

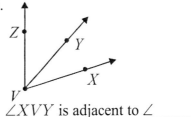

∠*XVY* is adjacent to ∠_____

3.

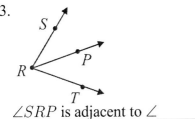

∠*SRP* is adjacent to ∠_____

7.

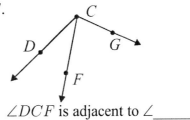

∠*DCF* is adjacent to ∠_____

4.

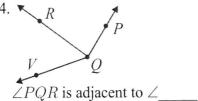

∠*PQR* is adjacent to ∠_____

8.

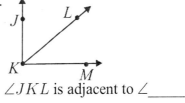

∠*JKL* is adjacent to ∠_____

14.3   Vertical Angles

When two lines intersect, two pairs of vertical angles are formed. Vertical angles are not adjacent. Vertical angles have the same measure.

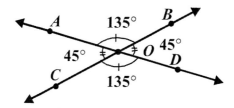

$\angle AOB$ and $\angle COD$ are vertical angles. $\angle AOC$ and $\angle BOD$ are vertical angles. **Vertical angles** are **congruent**. Congruent means they have the same measure.

In the diagram below, name the second angle in each pair of vertical angles.

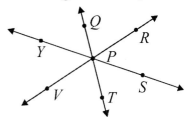   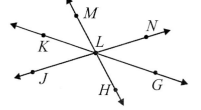

1. $\angle YPV$ _____   4. $\angle VPT$ _____   7. $\angle MLN$ _____   10. $\angle GLM$ _____
2. $\angle QPR$ _____   5. $\angle RPT$ _____   8. $\angle KLH$ _____   11. $\angle KLM$ _____
3. $\angle SPT$ _____   6. $\angle VPS$ _____   9. $\angle GLN$ _____   12. $\angle HLG$ _____

Use the information given to find the measure of each unknown vertical angle.

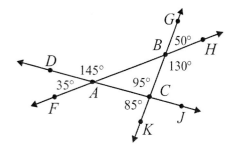   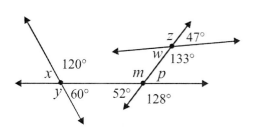

13. $\angle CAF =$ _____   19. $\angle x =$ _____

14. $\angle ABC =$ _____   20. $\angle y =$ _____

15. $\angle KCJ =$ _____   21. $\angle z =$ _____

16. $\angle ABG =$ _____   22. $\angle w =$ _____

17. $\angle BCJ =$ _____   23. $\angle m =$ _____

18. $\angle CAB =$ _____   24. $\angle p =$ _____

14.4 Complementary and Supplementary Angles

Two angles are **complementary** if the sum of the measures of the angles is 90°.

Two angles are **supplementary** if the sum of the measures of the angles is 180°. A **linear pair** is a pair of adjacent angles that are supplementary. Below the angles 32° and 148° are a linear pair. The angles may be adjacent but do not need to be.

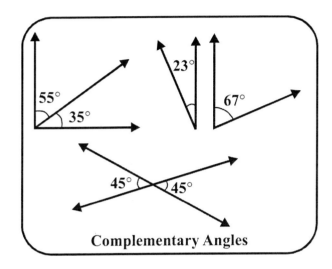

Complementary Angles

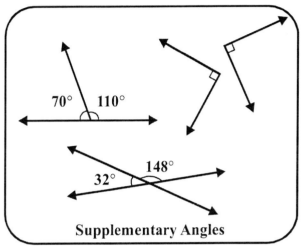

Supplementary Angles

Calculate the measure of each unknown angle.

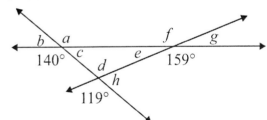

1. ∠a = _____
2. ∠b = _____
3. ∠c = _____
4. ∠d = _____

5. ∠e = _____
6. ∠f = _____
7. ∠g = _____
8. ∠h = _____

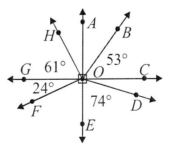

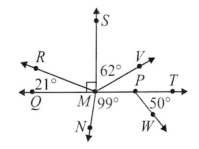

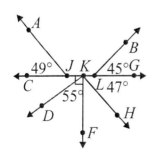

9. ∠AOB = _____

10. ∠COD = _____

11. ∠EOF = _____

12. ∠AOH = _____

13. ∠RMS = _____

14. ∠VMT = _____

15. ∠QMN = _____

16. ∠WPQ = _____

17. ∠AJK = _____

18. ∠CKD = _____

19. ∠FKH = _____

20. ∠BLC = _____

14.5   Corresponding, Alternate Interior, and Alternate Exterior Angles

If two parallel lines are intersected by a **transversal**, a line passing through both parallel lines, the **corresponding angles** are congruent.

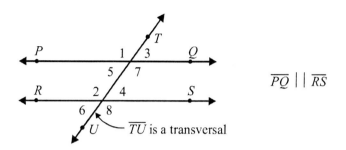

$\overline{PQ} \mid\mid \overline{RS}$

$\overline{TU}$ is a transversal

∠1 and ∠2 are congruent.   They are corresponding angles.
∠3 and ∠4 are congruent.   They are corresponding angles.
∠5 and ∠6 are congruent.   They are corresponding angles.
∠7 and ∠8 are congruent.   They are corresponding angles.

Alternate interior angles are also congruent. They are on the opposite sides of the transversal and inside the parallel lines.

∠5 and ∠4 are congruent.   They are alternate interior angles.
∠7 and ∠2 are congruent.   They are alternate interior angles.

Alternate exterior angles are also congruent. They are on the opposite sides of the transversal and above and below the parallel lines.

∠1 and ∠8 are congruent.   They are alternate exterior angles.
∠3 and ∠6 are congruent.   They are alternate exterior angles.

Look at the diagram below. For each pair of angles, state whether they are corresponding (C), alternate interior (I), alternate exterior (E), vertical (V), or supplementary angles (S).

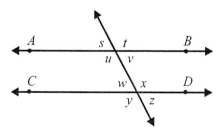

$\overline{AB}$ and $\overline{CD}$ are parallel.

1. ∠u, ∠x

2. ∠w, ∠s

3. ∠t, ∠y

4. ∠s, ∠t

5. ∠w, ∠y

6. ∠t, ∠x

7. ∠w, ∠z

8. ∠v, ∠w

9. ∠v, ∠z

10. ∠s, ∠z

11. ∠t, ∠u

12. ∠w, ∠x

13. ∠w, ∠s

14. ∠s, ∠v

15. ∠x, ∠z

14.6 Sum of Interior Angles of a Polygon

Given a convex polygon, you can find the sum of the measures of the interior angles using the following formula: Sum of the measures of the interior angles $= 180° (n - 2)$, where n is the number of sides of the polygon.

Example 1:   Find the sum of the measures of the interior angles of the following polygon:

Solution: The figure has 8 sides. Using the formula we have $180° (8 - 2) = 180° (6) = 1080°$

Using the formula, $180° (n - 2)$, find the sum of the interior angles of the following figures.

1.

4.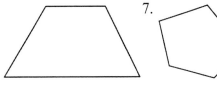

7.

10.

2.

5.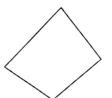

8.

11.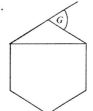

3.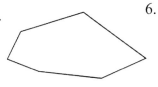

6.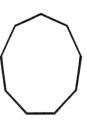

9.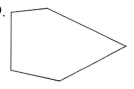

12.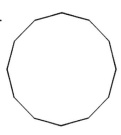

Find the measure of $\angle G$ in the regular polygons shown below. Remember that the sides of a regular polygon are equal.

13.

14.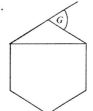

15.

Chapter 14 Review

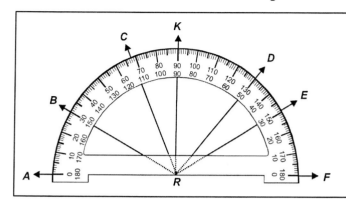

1. What is the measure of ∠*DRA*?

2. What is the measure of ∠*CRF*?

3. What is the measure of ∠*ARB*?

Use the following diagram for questions 4–14.

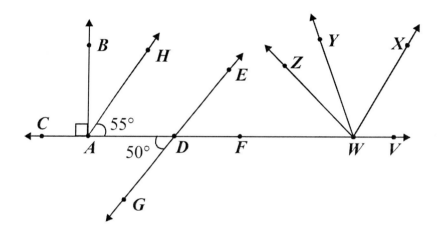

4. Which angle is a supplementary angle to ∠*EDF*?

5. What is the measure of ∠*GDF*?

6. Which two angles are right angles?

7. What is the measure of ∠*EDF*?

8. Which angle is adjacent to ∠*BAD*?

9. Which angle is a complementary angle to ∠*HAD*?

10. What is the measure of ∠*HAB*?

11. What is the measure of ∠*CAD*?

12. What kind of angle is ∠*FDA*?

13. What kind of angle is ∠*GDA*?

14. Which angles are adjacent to ∠*EDA*?

Look at the diagram below. For each pair of angles, state whether they are corresponding (C), alternate interior (I), alternate exterior (E), vertical (V), or supplementary (S) angles.

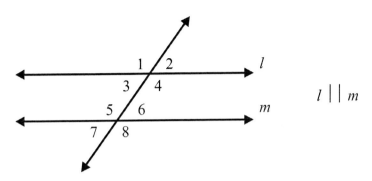

$l \parallel m$

15. $\angle 1$ and $\angle 4$

16. $\angle 2$ and $\angle 6$

17. $\angle 1$ and $\angle 3$

18. $\angle 5$ and $\angle 8$

19. $\angle 5$ and $\angle 7$

20. $\angle 6$ and $\angle 5$

21. $\angle 2$ and $\angle 7$

22. $\angle 1$ and $\angle 2$

23. $\angle 4$ and $\angle 5$

24. $\angle 6$ and $\angle 8$

25. $\angle 3$ and $\angle 6$

26. $\angle 4$ and $\angle 8$

27. $\angle 1$ and $\angle 5$

28. $\angle 2$ and $\angle 3$

29. What is the sum of the measures of the interior angles in the figure below?

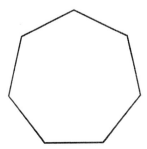

Chapter 14 Test

1. What type of angle is shown below?

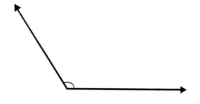

A right
B acute
C obtuse
D straight

2. What is the sum of two complementary angles?

A 180°
B 45°
C 90°
D 360°

3. What is the measure of an angle that is supplementary to 87°?

A −42°
B 3°
C 273°
D 93°

4. In the diagram below, which two angles form a linear pair?

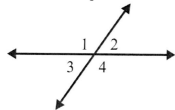

A ∠1 and ∠2
B ∠1 and ∠3
C ∠1 and ∠4
D Both A and B are correct.

Use the following diagram to answer questions 5–8.

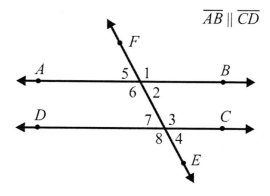

$\overline{AB} \parallel \overline{CD}$

5. Which angles are vertical angles?

A ∠1 and ∠2
B ∠1 and ∠3
C ∠1 and ∠4
D ∠1 and ∠6

6. Which angles are alternate exterior angles?

A ∠2 and ∠7
B ∠3 and ∠8
C ∠1 and ∠8
D ∠5 and ∠3

7. Which angles are alternate interior angles?

A ∠7 and ∠8
B ∠6 and ∠3
C ∠2 and ∠3
D ∠1 and ∠4

8. Which angles are corresponding angles?

A ∠1 and ∠3
B ∠1 and ∠4
C ∠7 and ∠6
D ∠1 and ∠8

Chapter 15
Triangles

This chapter covers the following Alabama objectives and standards in mathematics:

	Objective(s)
Standard VII	1, 2, 3

15.1   Types of Triangles

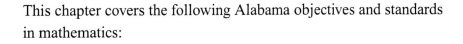

right triangle
contains 1 right $\angle$

acute triangle
all angles are acute
(less than 90°)

obtuse triangle
one angle is obtuse
(greater than 90°)

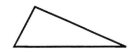

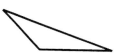

equilateral triangle
all three sides equal
all angles are 60°

scalene triangle
no sides equal
no angles equal

isosceles triangle
two sides equal
two angles equal

15.2   Interior Angles of a Triangle

The three interior angles of a triangle always add up to $180°$.

Example 1:

$45° + 45° + 90° = 180°$

$30° + 60° + 90° = 180°$

$60° + 60° + 60° = 180°$

Example 2:   Find the missing angle in the triangle.

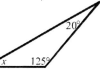

Solution:

$$
\begin{aligned}
20° + 125° + x &= 180° \\
-20° \;\; -125° & \quad\quad -20° \;\; -125° \\
x &= 180° - 20° - 125° \\
x &= 35°
\end{aligned}
$$

Subtract $20°$ and $125°$ from both sides to get x by itself.

The missing angle is $35°$.

Find the missing angle in the triangles.

1.

2.
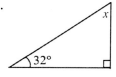

3.
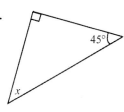

4.
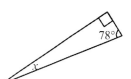

5.
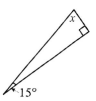

6.
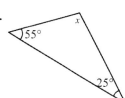

7.
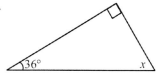

8.
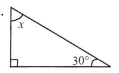

9.
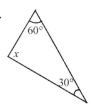

Find the missing angles in the triangles.

10.
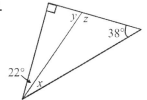

11.
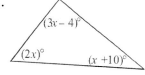

12.
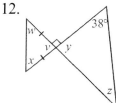

15.3   Exterior Angles

The **exterior angle** of a triangle is always equal to the sum of the opposite interior angles.

Example 3:   Find the measure of $\angle x$ and $\angle y$.

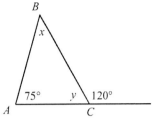

Step 1:   Using the rule for exterior angles,
$120° = \angle A + \angle B$
$120° = 75° + x$
$120° - 75° = 75° - 75° + x$
$45° = x$

Step 2:   The sum of the interior angles of a triangle equals 180°, so
$180° = 75° + 45° + y$
$180° - 75° - 45° = 75° - 75° + 45° - 45° + y$
$60° = y$

Find the measures of x and y.

1.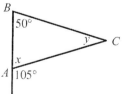

3.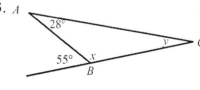

5.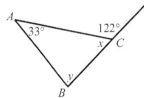

2.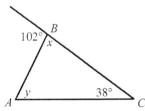

4.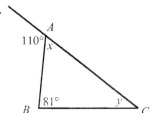

6.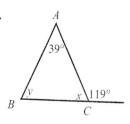

Find the measures of the angles.

7.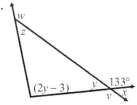

8.

9.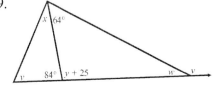

15.4   Triangle Inequality Theorem

The triangle inequality theorem states that the sum of the measure of any two sides in a triangle must be greater than the measure of the third side.

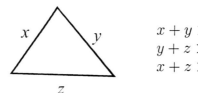

$$x + y > z$$
$$y + z > x$$
$$x + z > y$$

Example 4:   Determine whether or not it is possible to create a triangle with sides of 1 units, 5 units, and 7 units.

Step 1:   First, you must set up three inequalities. Remember the sum of any two sides of a triangle must be greater than the third side.

$$1 + 5 > 7 \qquad 1 + 7 > 5 \qquad 5 + 7 > 1$$

Step 2:   Determine if the inequalities are true.

$$
\begin{array}{ccc}
1 + 5 > 7 & 1 + 7 > 5 & 5 + 7 > 1 \\
6 > 7 & 8 > 5 & 12 > 1 \\
\text{False} & \text{True} & \text{True}
\end{array}
$$

The number 6 is not greater than 7, so a triangle cannot be formed using the sides given.
(All three inequalities must be true in order to create a triangle.)

Determine whether or not it is possible to create a triangle given the following measures of sides. Write "yes" if it is possible to form a triangle with the given measures of sides or write "no" if it is not possible.

1. 7, 8, 13
2. 2, 5, 9
3. 10, 8, 15

4. 6, 9, 20
5. 101, 89, 150
6. 1, 2, 4

7. 7, 7, 14
8. 21, 15, 29
9. 11, 9, 17

15.5 Similar Triangles

Two triangles are similar if the measurements of the three angles in both triangles are the same. If the three angles are the same, then their corresponding sides are proportional.

Corresponding Sides - The triangles below are similar. Therefore, the two shortest sides from each triangle, c and f, are corresponding. The two longest sides from each triangle, a and d, are corresponding. The two medium length sides, b and e, are corresponding.

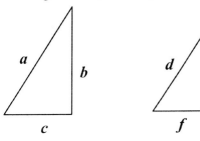

Proportional - The corresponding sides of similar triangles are proportional to each other. This means if we know all the measurements of one triangle, and we know only one measurement of the other triangle, we can figure out the measurements of the two other sides with proportion problems. The two triangles below are similar.

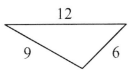 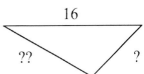

Note: To set up the proportion correctly, it is important to keep the measurements of each triangle on opposite sides of the equal sign.

To find the short side:
Step 1: Set up the proportion

$$\frac{\text{long side}}{\text{short side}} \quad \frac{12}{6} = \frac{16}{?}$$

Step 2: Solve the proportion. Multiply the two numbers diagonal to each other and then divide by the other number.
$$16 \times 6 = 96$$
$$96 \div 12 = 8$$

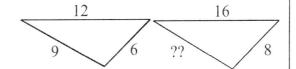

To find the medium length side:
Step 1: Set up the proportion

$$\frac{\text{long side}}{\text{medium}} \quad \frac{12}{9} = \frac{16}{??}$$

Step 2: Solve the proportion. Multiply the two numbers diagonal to each other and then divide by the other number.
$$16 \times 9 = 144$$
$$144 \div 12 = 12$$

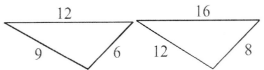

To find the scale factor in the problem on the previous page, we must divide a value from the second triangle by the corresponding value from the first triangle. The value 16 is from the second triangle, and the corresponding value from the first triangle is 12. $k = \dfrac{16}{12} = \dfrac{4}{3}$

The scale factor in this problem is $\frac{4}{3}$.

To check this answer multiply every term in the first triangle by the scale factor, and you will get every term in the second triangle.

$$12 \times \frac{4}{3} = 16 \qquad 9 \times \frac{4}{3} = 12 \qquad 6 \times \frac{4}{3} = 8$$

Find the missing side from the following similar triangles.

1.

5.

2.

6.

3.

7.

4.

8.

15.6   Pythagorean Theorem

Pythagoras was a Greek mathematician and philosopher who lived around 600 B.C. He started a math club among Greek aristocrats called the Pythagoreans. Pythagoras formulated the **Pythagorean Theorem** which states that in a **right triangle**, the sum of the squares of the legs of the triangle are equal to the square of the hypotenuse. Most often you will see this formula written as $a^2 + b^2 = c^2$. **This relationship is only true for right triangles.**

Example 5:   Find the length of side c.

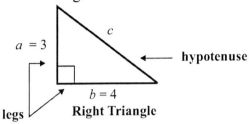

Formula: $a^2 + b^2 = c^2$
$3^2 + 4^2 = c^2$
$9 + 16 = c^2$
$25 = c^2$
$\sqrt{25} = \sqrt{c^2}$
$5 = c$

Find the hypotenuse of the following triangles. Round the answers to two decimal places.

1.

$c =$ _____

4.

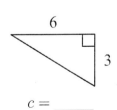

$c =$ _____

7.

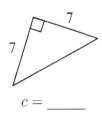

$c =$ _____

2.

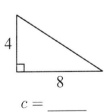

$c =$ _____

5.

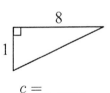

$c =$ _____

8.

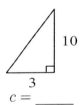

$c =$ _____

3.

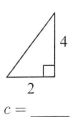

$c =$ _____

6.

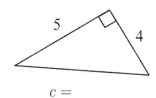

$c =$ _____

9.

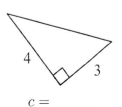

$c =$ _____

15.7 Finding the Missing Leg of a Right Triangle

In some triangles, we know the measurement of the hypotenuse as well as one of the legs. To find the measurement of the other leg, use the Pythagorean theorem by filling in the known measurements, and then solve for the unknown side.

Example 6: Find the measure of b.

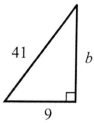

In the formula, $a^2 + b^2 = c^2$, a and b are the legs and c is always the hypotenuse.

$9^2 + b^2 = 41^2$
$81 + b^2 = 1681$
$b^2 = 1681 - 81$
$b^2 = 1600$
$\sqrt{b^2} = \sqrt{1600}$
$b = 40$

Practice finding the measure of the missing leg in each right triangle below. Simplify square roots.

1.
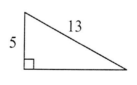

_____

4.
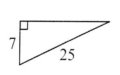

_____

7.

_____

2.
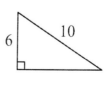

_____

5.
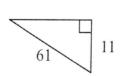

_____

8.
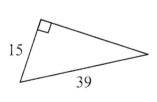

_____

3.
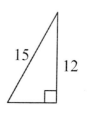

_____

6.
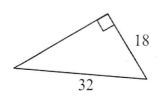

_____

9.
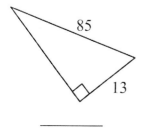

_____

15.8   Applications of the Pythagorean Theorem

The Pythagorean Theorem can be used to determine the distance between two points in some situations. Recall that the formula is written $a^2 + b^2 = c^2$.

Example 7:   Find the distance between point B and point A given that the length of each square is 1 inch long and 1 inch wide.

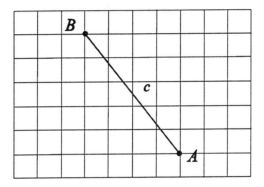

Step 1:   Draw a straight line between the two points. We will call this side c.

Step 2:   Draw two more lines, one from point B and one from point A. These lines should make a 90° angle. The two new lines will be labeled a and b. Now we can use the Pythagorean Theorem to find the distance from Point B to Point A.

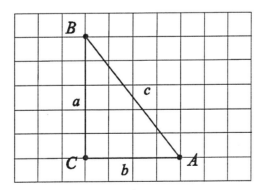

Step 3:   Find the length of a and c by counting the number of squares each line has. We find that $a = 5$ inches and $b = 4$ inches. Now, substitute the values found into the Pythagorean Theorem.

$$
\begin{aligned}
a^2 + b^2 &= c^2 \\
5^2 + 4^2 &= c^2 \\
25 + 16 &= c^2 \\
41 &= c^2 \\
\sqrt{41} &= \sqrt{c^2} \\
\sqrt{41} &= c
\end{aligned}
$$

Use the Pythagorean Theorem to find the distances asked. Round your answers to two decimal points.

Below is a diagram of the mall. Use the grid to help answer questions 1 and 2. Each square is 25 feet × 25 feet.

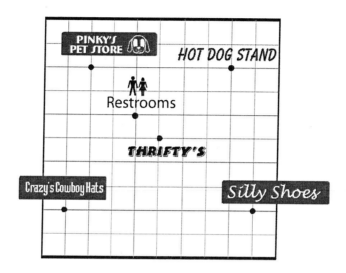

1. Marty walks from Pinky's Pet Store to the restroom to wash his hands. How far did he walk?

2. Betty needs to meet her friend at Silly Shoes, but she wants to get a hot dog first. If Betty is at Thrifty's, how far will she walk to meet her friend?

Below is a diagram of a football field. Use the grid on the football field to help find the answers to questions 3 and 4. Each square is 10 yards × 10 yards.

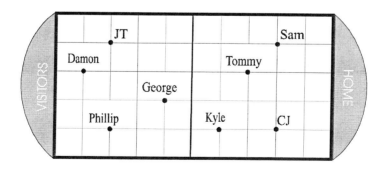

3. George must throw the football to a teammate before he is tackled. If CJ is the only person open, how far must George be able to throw the ball?

4. Damon has the football and is close to scoring a touchdown. If Phillip tries to stop him, how far must he run to reach Damon?

Chapter 15 Review

1. Find the missing angle.

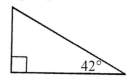

2. What is the length of line segment $\overline{WY}$?

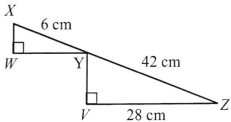

3. Find the missing side.

4. Find the measure of the missing leg of the right triangle below.

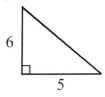

5. The following two triangles are similar. Find the length of the missing side.

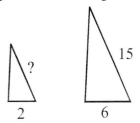

For questions 6–7, determine if the measures of sides given can form a triangle.

6. 1, 5, 3

7. 16, 22, 31

8.

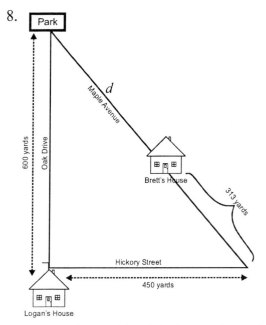

Logan enjoys taking his dog to the park. Some days he leaves his house, located on the corner of Hickory St. and Oak Dr., and walks directly to the park. Sometimes, though, he walks down Hickory St., turns onto Maple Ave. to meet his friend, Brett, and then continues on Maple Ave. to the park. What is the approximate distance (d) from Brett's house to the park?

For questions 9–10, find the missing angles.

9.

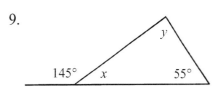

10.

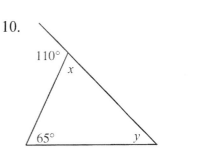

Chapter 15 Test

1. What is the measure of missing angle?

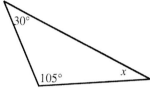

   A 225°

   B 45°

   C 75°

   D 30°

2. What is the measure of y?

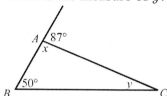

   A 93°

   B 87°

   C 37°

   D Cannot be determined

3. What is the measure of the missing side in the triangle?

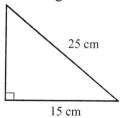

   A 5 cm

   B 29 cm

   C 10 cm

   D 20 cm

4. Given the measures of sides below, which cannot form a triangle?

   A 5, 7, 10

   B 2, 3, 4

   C 15, 6, 9

   D 19, 20, 36

5. What type of triangle is illustrated?

   A right

   B isosceles

   C equilateral

   D obtuse

6. Approximately what is the measure of the hypotenuse of the triangle?

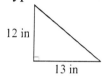

   A 14 in

   B 157 in

   C 18 in

   D 313 in

7. What is the measure of the two missing angles?

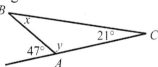

   A $x = 26°, y = 133°$

   B $x = 47°, y = 112°$

   C $x = 112°, y = 47°$

   D $x = 133°, y = 26°$

8. What is the length of the line segment $\overline{WY}$?

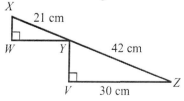

   A 15 cm

   B 16 cm

   C 18 cm

   D 30 cm

Chapter 16
Plane Geometry

This chapter covers the following Alabama objectives and standards in mathematics:

	Objective(s)
Standard IV	1
Standard VII	3, 4

16.1   Types of Polygons

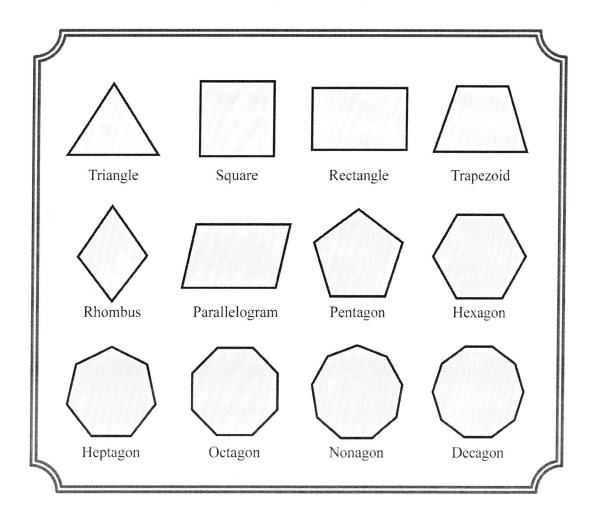

16.2   Perimeter

The **perimeter** is the distance around a polygon. To find the perimeter, add the lengths of the sides.

Examples:

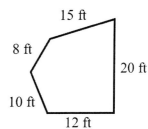

$P = 7 + 15 + 7 + 15$
$P = 44 \, \text{in}$

$P = 4 + 6 + 5$
$P = 15 \, \text{cm}$

$P = 8 + 15 + 20 + 12 + 10$
$P = 65 \, \text{ft}$

Find the perimeter of the following polygons.

1.

5.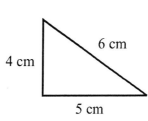

9.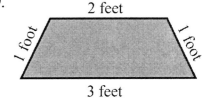

2.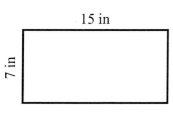

6.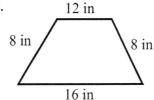

10. a regular pentagon with sides of 12 centimeters

11. a regular square with sides of 7 inches

3.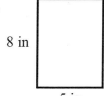

7.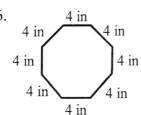

12. a regular decagon with sides of 5 centimeters

12. a pentagon with sides of 2 feet

4.

8.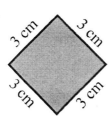

13. a triangle with 3 equal sides, 4 inches long

14. a regular hexagon with sides of 5 centimeters

16.3   Area of Squares and Rectangles

Area - area is always expressed in square units, such as in^2, m^2, and ft^2.

The area, (A), of squares and rectangles equals length (l) times width (w). $A = l \times w$.

Example 1:

4 cm

4 cm

$A = lw$
$A = 4 \times 4$
$A = 16 \ cm^2$

If a square has an area of $16 \ cm^2$, it means that it will take 16 squares that are 1 cm on each side to cover the area that is 4 cm on each side.

Find the area of the following squares and rectangles using the formula $A = lw$.

1.
10 ft
10 ft

2.
5 cm
2 cm

3.
4 in
9 in

4.
9 in
20 in

5.
6 ft
6 ft

6.
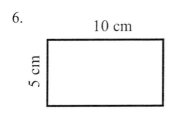
10 cm
5 cm

7.
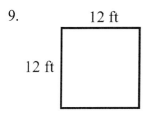
4 ft
2 ft

8.
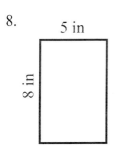
5 in
8 in

9.
12 ft
12 ft

10.
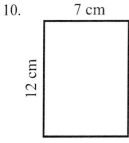
7 cm
12 cm

11.
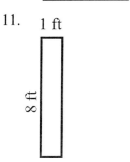
1 ft
8 ft

12.
6 cm
7 cm

16.4   Area of Triangles

Example 2:   Find the area of the following triangle.
The formula for the area of a triangle is as follows:

$$A = \frac{1}{2} \times b \times h$$

A = area
b = base
h = height or altitude

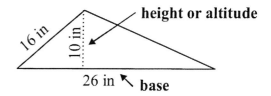

Step 1:   Insert the measurements from the triangle into the formula: $A = \frac{1}{2} \times 26 \times 10$

Step 2:   Cancel and multiply. $A = \frac{1}{\cancel{26}_{1}} \times \frac{\cancel{26}^{13}}{1} \times \frac{10}{1} = 130 \text{ in}^2$

Note: **Area is always expressed in square units such as in^2, ft^2, or m^2.**

Find the area of the following triangles. Remember to include units.

1.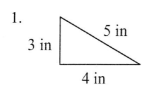

2.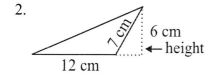

3.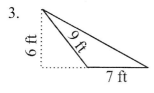

4.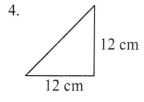

5.

6.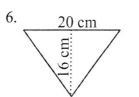

7.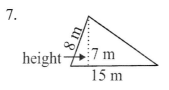

8.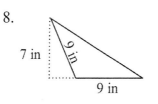

9.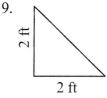

10.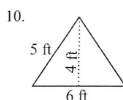

11.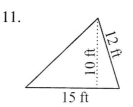

12.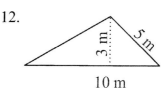

16.5   Area of Trapezoids and Parallelograms

Example 3:   Find the area of the following parallelogram.

The formula for the area of a parallelogram is $A = bh$.
A = area
b = base
h = height

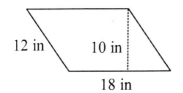

Step 1:   Insert measurements from the parallelogram into the formula: $A = 18 \times 10$.

Step 2:   Multiply. $18 \times 10 = 180$ in^2

Example 4:   Find the area of the following trapezoid.
The formula for the area of a trapezoid is $A = \frac{1}{2}h(b_1 + b_2)$. A trapezoid has two bases that are parallel to each other. When you add the length of the two bases together and then multiply by $\frac{1}{2}$, you find their average length.

A = area
b = base
h = height

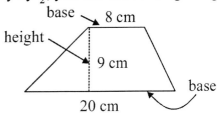

Insert the measurements from the trapezoid into the formula and solve:
$\frac{1}{2} \times 9\,(8 + 20) = 126$ cm^2

Find the area of the following parallelograms and trapezoids.

1.
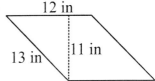

4.
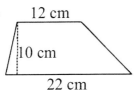

7.
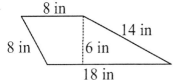

2.
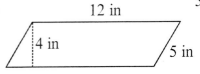

5.
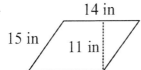

8.
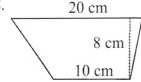

3.
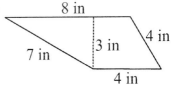

6.
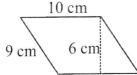

9.
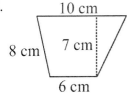

16.6   Circumference

Circumference, C - the distance around the outside of a circle
Diameter, d - a line segment passing through the center of a circle from one side to the other
Radius, r - a line segment from the center of a circle to the edge of a circle
Pi, π- the ratio of a circumference of a circle to its diameter $\pi = 3.14$ or $\pi = \frac{22}{7}$

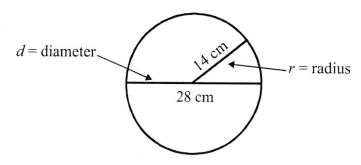

The formula for the circumference of a circle is $C = 2\pi r$ or $C = \pi d$. (The formulas are equal because the diameter is equal to twice the radius, $d = 2r$.)

Example 5:   Find the circumference of the circle above.

$C = \pi d$ Use $\pi = 3.14$ $C = 2\pi r$
$C = 3.14 \times 28$ $C = 2 \times 3.14 \times 14$
$C = 87.92\,\text{cm}$ $C = 87.92\,\text{cm}$

Use the formulas given above to find the circumferences of the following circles. Use $\pi = 3.14$.

1. 8 in   2. 14 ft   3. 2 cm   4. 6 m   5. 8 ft

$C =$ _____ $C =$ _____ $C =$ _____ $C =$ _____ $C =$ _____

Use the formulas given above to find the circumferences of the following circles. Use $\pi = \frac{22}{7}$.

6. 3 ft   7. 12 in   8. 6 m   9. 5 cm   10. 16 in

$C =$ _____ $C =$ _____ $C =$ _____ $C =$ _____ $C =$ _____

16.7   Area of a Circle

The formula for the area of a circle is $A = \pi r^2$. The area is how many square units of measure would fit inside a circle.

Example 6:   Find the area of the circle, using both values for π.

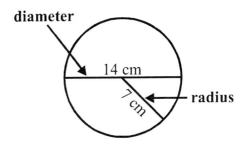

$$\text{Let } \pi = \frac{22}{7}$$
$$A = \pi r^2$$
$$A = \frac{22}{7} \times 7^2$$
$$A = \frac{22}{7}^1 \times \frac{\cancel{49}\ 7}{1}$$
$$= 154 \text{ cm}^2$$

$$\text{Let } \pi = 3.14$$
$$A = \pi r^2$$
$$A = 3.14 \times 7^2$$
$$A = 3.14 \times 49$$
$$= 153.86 \text{ cm}^2 \approx 154 \text{ cm}^2$$

Find the area of the following circles. Remember to include units.

Fill in the chart below. Include appropriate units.

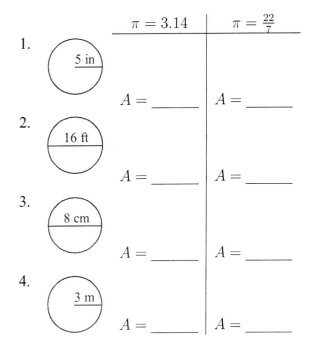

1. 5 in

$A =$ _____ ($\pi = 3.14$) $A =$ _____ ($\pi = \frac{22}{7}$)

2. 16 ft

$A =$ _____ $A =$ _____

3. 8 cm

$A =$ _____ $A =$ _____

4. 3 m

$A =$ _____ $A =$ _____

	Radius	Diameter	Area $\pi = 3.14$	Area $\pi = \frac{22}{7}$
5.	9 ft			
6.		4 in		
7.	8 cm			
8.		20 ft		
9.	14 m			
10.		18 cm		
11.	12 ft			
12.		6 in		

16.8   Two-Step Area Problems

Solving the problems below will require two steps. You will need to find the area of two figures, and then either add or subtract the two areas to find the answer. **Carefully read the examples.**

Example 7:
Find the area of the living room below.
Figure 1

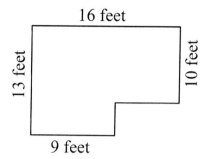

Step 1: Complete the rectangle as in Figure 2, and compute the area as if it were a complete rectangle.

Figure 2

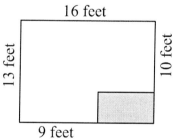

$$A = \text{length} \times \text{width}$$
$$A = 16 \times 13$$
$$A = 208 \text{ ft}^2$$

Step 2: Figure the area of the shaded part.

7 feet

3 feet

$$7 \times 3 = 21 \text{ ft}^2$$

Step 3: Subtract the area of the shaded part from the area of the complete rectangle

$$208 - 21 = 187 \text{ ft}^2$$

Example 8:
Find the area of the shaded sidewalk.

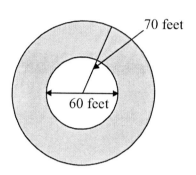

Step 1: Find the area of the outside circle.
$$\pi = 3.14$$
$$A = 3.14 \times 70 \times 70$$
$$A = 15,386 \text{ ft}^2$$

Step 2: Find the area of the inside circle.
$$\pi = 3.14$$
$$A = 3.14 \times 30 \times 30$$
$$A = 2826 \text{ ft}^2$$

Step 3: Subtract the area of the inside circle from the area of the outside circle.
$$15,386 - 2826 = 12,560 \text{ ft}^2$$

Find the area of the following figures.

1.

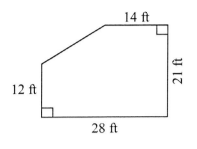

2.

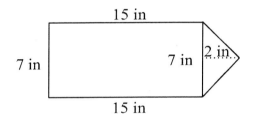

3. What is the area of the shaded circle? Use $\pi = 3.14$, and round the answer to the nearest whole number.

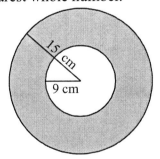

4.

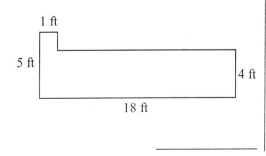

5. What is the area of the shaded part?

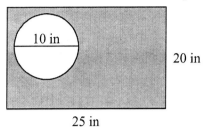

6. What is the area of the shaded part?

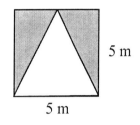

7. What is the area of the shaded part?

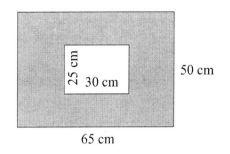

8.

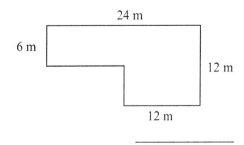

16.9   Similar Figures

The measures of corresponding sides of similar figures can also be found by setting up a proportion.

Example 9:   The following rectangles are similar. Find the value of x.

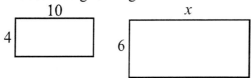

Step 1:   Set up the proportion.

$$\frac{\text{Short side rectangle 1}}{\text{Long side rectangle 1}} = \frac{\text{Short side rectangle 2}}{\text{Long side rectangle 2}}$$

Step 2: $\frac{4}{10} = \frac{6}{x}$ Cross multiply.

$4x = 60$

$x = 15$

All of the pairs of figures below are similar. Find the missing side for each pair.

1.

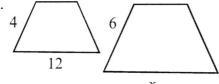

2.

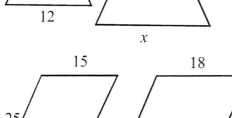

3.

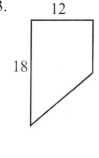

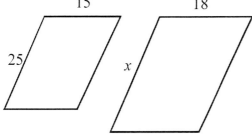

4.

5.

6.

16.10   Plane Geometry Word Problems

Solve the following word problems using the area, perimeter, and similar figure skills you have learned in this chapter.

1. Marvin has a piece of sheet metal measuring 3 feet by 2 feet. Marvin needs to remove a section on one end measuring $\frac{3}{4}$ of a foot by 2 feet. How much will remain of the sheet of metal?

2. Skeeter is helping with the scenery for the school play. He needs to paint a building on a sheet of plywood and show some sky at the top of the plywood. The sheet of plywood measures 4 feet by 8 feet. The sky painted at the top measure 4 feet by 1 foot. How much of the plywood is painted as a building?

3. Jose's dad is laying cement in a seating area of the garden. His dad wants to leave a flower bed area in the middle measuring 2 feet by 8 feet. The whole area measures 225 square feet. How many square feet of surface area will Jose's dad lay cement?

4. Mary is cutting out fabric to make a wall hanging for her room. The fabric measures 4 feet by 6 feet. She only needs 2 feet by 3 feet. How many square feet of fabric will Mary have leftover?

5. Karissa is drawing a garden design for her backyard. She has a plot 20 feet by 30 feet she will divide into 3 parts. The first part is a 15 feet by 15 feet, the second is a rectangle with an area of 175 ft^2. What is the area of the 3rd part?

6. Jeremy has cut a triangle with a base of 3 feet and a height of $1\frac{1}{2}$ feet out of a 4 foot square of plywood. Jeremy is planning on painting a design on the triangle and then hanging it in the basement to decorate the family room. How many square feet of plywood will Jeremy have left over?

7. Mr. Landish is trying to figure out the square footage of his lawn so he can buy the appropriate amount of fertilizer. He knows his lot is 110 feet by 220 feet, his house is 1,600 ft^2, and his driveway is 550 ft^2. What is the area of Mr. Landish's lawn?

8. Betty and Jake are figuring the amount of carpet needed to cover their daughter's bedroom. The three girls share a room measuring 13 feet by 12 feet, and there are two closets they want to carpet measuring 3 feet by 6 feet and the other is 3 feet by 3 feet. What is the total number of square feet of carpet needed to cover the floor and both closets?

9. The perimeter of a cage for the opossum exhibit at a local zoo is 60 feet. The length of the cage is 20 feet. What is the area of the base of the cage?

10. Ching Ngo is making a pair of drapes for her living room. Each of the four drapery panels, when completed, measures 8 feet long by 4 feet wide. What is the total area of the four drapery panels?

11. A building that is 200 feet tall, casts a shadow at 4:00 pm of 300 feet. A tree beside the building is 40 feet tall. How long is the shadow of the tree?

12. A painting includes an equilateral triangle measuring 8 cm on all sides. Next to it is a circle with a diameter measuring 8 cm. Are the two figures similar?

13. A rectangle measures 12 cm by 8 cm. A similar rectangle measures 9 cm on its longer side. What is the perimeter of the smaller rectangle?

14. Mr. Williams has an architectural drawing of the home he plans to build. His daughter is concerned about the size of the bedroom. The architectural drawing says each quarter inch of drawing equals one foot of floor space (when the house is built). If the drawing for her bedroom is 3 inches by 4 inches, what will the floor measure when the home is built?

15. $\triangle GHI$ is similar to $\triangle XYZ$. If $\angle G \sim \angle X$ and $\angle H \sim \angle Y$, is $\angle I \sim \angle Z$?

16. A round table has a diameter of 60 inches. If Rhonda wants to make a round doily for the center of the table that is 20% of the size of the table, what will be the diameter of the doily?

16.11   Perimeter and Area with Algebraic Expressions

You have already calculated the perimeter and area of various shapes with given measurements. You must also understand how to find the perimeter of shapes that are described by algebraic expressions. Study the examples below.

Example 10:   Use the equation $P = 2l + 2w$ to find the perimeter of the following rectangle.

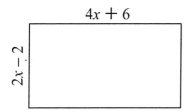

Step 1: Find $2l$.	$2(4x + 6) = 8x + 12$
Step 2: Find $2w$.	$2(2x - 2) = 4x - 4$
Step 3: Find $2l + 2w$.	$12x + 8$

Perimeter $= 12x + 8$

Example 11:   Using the formula $A = lw$, find the area of the rectangle below.

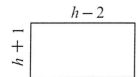

Step 1: $A = (h - 2)(h + 1)$
Step 2: $A = h^2 - 2h + h - 2$
Step 3: $A = h^2 - h - 2$

Area $= h^2 - h - 2$

Example 12:   Find the area of a circle with $r = 4x + 1$.

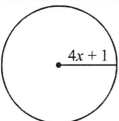

Step 1:   Substitute the radius $(r = 4x + 1)$ into the formula for a circle $A = \pi r^2$.
$A = \pi r^2 = \pi (4x + 1)^2$

Step 2:   Simplify.
$A = \pi (4x + 1)^2 = \pi (4x + 1)(4x + 1) = \pi (16x^2 + 8x + 1)$
$A = \pi (16x^2 + 8x + 1)$

Find the perimeter of each of the following figures.

1.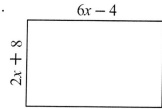
$6x - 4$
$2x + 8$

2. $4x + 3$
$4x - 5$

3.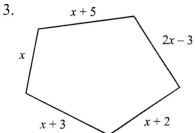
$x + 5$
$2x - 3$
x
$x + 3$
$x + 2$

4.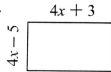
$4x + 3$
$3x + 2$

5. $6x + 3$
$4x + 2$

6.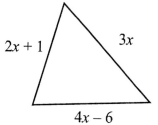
$2x + 1$
$3x$
$4x - 6$

Find the area of each of the following rectangles.

7. $5 - 2m$
$2 + m$

9. $2h - 2$
$2h - 2$

11. n
$n + 8$

8. $8 - 4n$
$5 - n$

10. $9 - h$
$4 + 2h$

12. $7 + 2b$
$2 + b$

Find the area of the figures below.

13.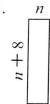
$2x - 6$

15. $5x + 3$

17. $3x$
$2x + 8$

14.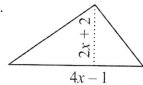
$2x + 2$
$4x - 1$

16.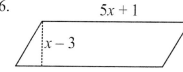
$5x + 1$
$x - 3$

18.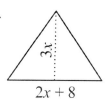
$2x + 1$
$x + 9$
$4x - 6$

Chapter 16 Review

1. Calculate the perimeter of the following figure.

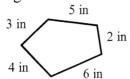

2. Find the area of the shaded region of the figure below.

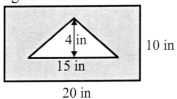

3. Calculate the perimeter and area.

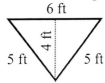

4. Calculate the perimeter and area.

7 in

4 in

5. Find the area.

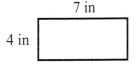

6. Find the area.

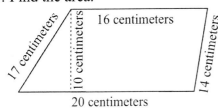

7. Find the area of the parallelogram.

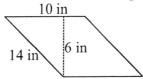

8. Find the missing side below.

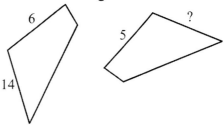

9. Find the area of the shaded part of the image below.

10. Calculate the circumference and the area of a circle with a radius $= 7$ cm. Use $\pi = \frac{22}{7}$.

11. Calculate the circumference and the area of a circle with a diameter $= 2$ ft. Use $\pi = 3.14$.

Use what you know about perimeter, circumference, and area to solve the following questions.

12. Find the perimeter and area of a rectangle with length $(x - 1)$ and width $(x + 6)$.

13. Find the circumference and area of a circle with radius equal to $3x - 2$.

14. Find the area of a triangle with height $(4x - 3)$ and base $(x + 2)$.

15. Find the perimeter and area of a rectangle with length $(3x - 8)$ and width $(x + 5)$.

Chapter 16 Test

1. Find the area.

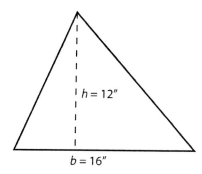

A 96 in
B 28 in²
C 96 in²
D 28 in

2. The figure below is a circle inscribed in a square. What is the area of the shaded region?

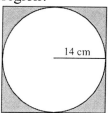

A 44 square centimeters
B 168 square centimeters
C 196 square centimeters
D 616 square centimeters

3. What is the area of the figure below?

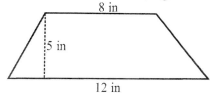

A 25 in²
B 50 in²
C 100 in²
D 480 in²

4. Using the formula $A = \frac{1}{2}bh$ for the area of a triangle, find the area of the triangle below.

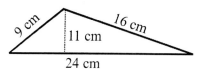

A 49 cm²
B 132 cm²
C 264 cm²
D 3,456 cm²

5. Find the area.

12″ (inches)

A 48 inches²
B 24 inches²
C 132 inches²
D 144 inches²

6. Find the area.

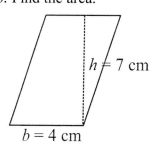

A 28 cm
B 28 cm²
C 22 cm
D 22 cm²

7. How many square feet of sod are needed to cover a 9-foot by 60-foot lawn?

   A 69 square feet

   B 138 square feet

   C 270 square feet

   D 540 square feet

8. Find the perimeter.

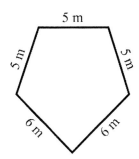

   A 30 m

   B 27 m

   C 25 m

   D 28 m

9. Find the perimeter.

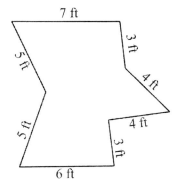

   A 40 ft

   B 38 ft

   C 37 ft

   D 36 ft

10. Find the area of the shaded region.

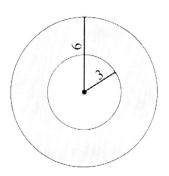

   A 24π

   B 3π

   C 27π

   D 15π

11. Find the area of the shaded region.

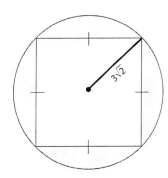

   A $3\sqrt{2}\pi - 36$

   B $18\pi - 36$

   C $18\pi - 18$

   D $3\sqrt{2}\pi - 18$

12. What is the area of a circle with a radius of 7 cm? (Round to the nearest whole number)

   A 154 square cm

   B 196 square cm

   C 347 square cm

   D 616 square cm

13. Which of the following figures is a parallelogram?

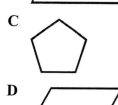

A

B

C

D

14. Find the circumference. Use $\pi = 3.14$.

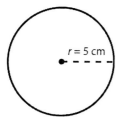

$r = 5$ cm

A 15.7 cm

B 62.8 cm

C 31.4 cm

D 0.314 cm

15. Find the area. Use $\pi = 3.14$.

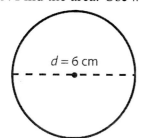

$d = 6$ cm

A 113.04 cm^2

B 28.26 cm^2

C 18.84 cm^2

D 188.4 cm^2

16. Which item below is not a polygon?

A triangle

B heptagon

C octagon

D circle

17. What is the name of the polygon below?

A quadrilateral

B pentagon

C hexagon

D octagon

18. What is the circumference of a circle that has a diameter of 10 cm?

A 15.7 cm

B 31.4 cm

C 78.5 cm

D 310 cm

19. What is the measure of x?

9

2

3

x

A 10

B 13.5

C 26

D Cannot be determined without the measure of the other sides.

20. The length of the rectangle is 5 units longer than the width. Which expression could be used to represent the area of the rectangle?

A $w^2 + 5w$

B $w^2 + 5$

C $w^2 + 25$

D $w^2 + 10w + 25$

Chapter 17
Solid Geometry

This chapter covers the following Alabama objectives and standards in mathematics:

	Objective(s)
Standard IV	1
Standard VII	4

17.1   Understanding Volume

Volume - Measurement of volume is expressed in cubic units such as in^3, ft^3, m^3, cm^3, or mm^3. The volume of a solid is the number of cubic units that can be contained in the solid.

First, let's look at rectangular solids.

Example 1:   How many 1 cubic centimeter cubes will it take to fill up the figure below?

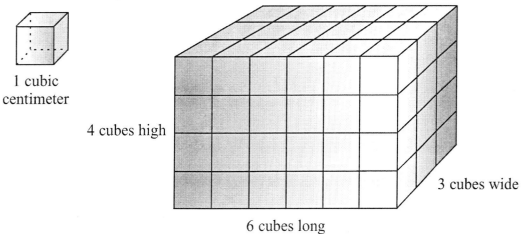

1 cubic centimeter

4 cubes high

3 cubes wide

6 cubes long

To find the volume, you need to multiply the length times the width times the height.
Volume of a rectangular solid = length $\times$ width $\times$ height $(V = lwh)$.

$$V = 6 \times 3 \times 4 = 72 \text{ cm}^3$$

17.2   Volume of Rectangular Prisms

You can calculate the volume (V) of a rectangular prism (box) by multiplying the length (l) by the width (w) by the height (h), as expressed in the formula $V = (lwh)$.

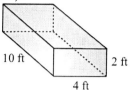

Example 2:   Find the volume of the box pictured here:

10 ft   2 ft   4 ft

Step 1:   Insert measurements from the figure into the formula.

Step 2:   Multiply to solve. $10 \times 4 \times 2 = 80 \text{ ft}^3$

Note: **Volume is always expressed in cubic units such as** in^3, ft^3, m^3, cm^3, **or** mm^3.

Find the volume of the following rectangular prisms (boxes).

1.
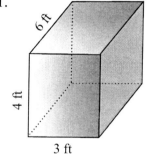
6 ft   4 ft   3 ft

4.
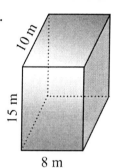
10 m   15 m   8 m

7.
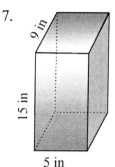
9 in   15 in   5 in

2.
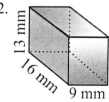
13 mm   16 mm   9 mm

5.
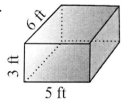
6 ft   3 ft   5 ft

8.
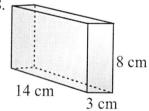
8 cm   14 cm   3 cm

3.
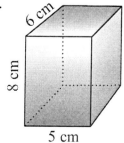
6 cm   8 cm   5 cm

6.
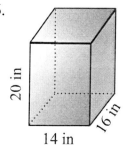
20 in   16 in   14 in

9.
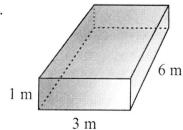
1 m   6 m   3 m

17.3   Volume of Cubes

A **cube** is a special kind of rectangular prism (box). Each side of a cube has the same measure. So, the formula for the volume of a cube is $V = s^3$ ($s \times s \times s$).

Example 3:   Find the volume of the cube at right:

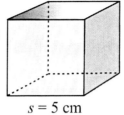

$s = 5$ cm

Step 1:   Insert measurements from the figure into the formula.

Step 2:   Multiply to solve. $5 \times 5 \times 5 = 125$ cm^3

Note: Volume is always expressed in cubic units such as in^3, ft^3, m^3, cm^3, **or** mm^3.

Answer each of the following questions about cubes.

1. If a cube is 3 centimeters on each edge, what is the volume of the cube?

2. If the measure of the edge is doubled to 6 centimeters on each edge, what is the volume of the cube?

3. If the edge of a 3-centimeter cube is tripled to become 9 centimeters on each edge, what will the volume be?

4. How many cubes with edges measuring 3 centimeters would you need to stack together to make a solid 12-centimeter cube?

5. What is the volume of a 2-centimeter cube?

6. Jerry built a 2-inch cube to hold his marble collection. He wants to build a cube with a volume 8-times larger. How much will each edge measure?

Find the volume of the following cubes.

7.

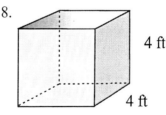

$s = 7$ in.

8.

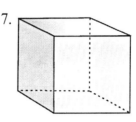

4 ft

4 ft

4 ft

9. 12 inches = 1 foot

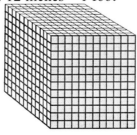

$s = 1$ foot

How many cubic inches are in a cubic foot?

17.4   Volume of Cylinders

To find the volume of a cylinder, insert the measurements given for the solid into the correct formula and solve. Remember, volumes are expressed in cubic units such as in^3, ft^3, m^3, cm^3, or mm^3. The volume of a cylinder is $V = \pi r^2 h$.

Example 4:   Find the volume of the cylinder below.

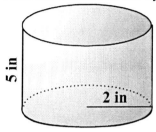

Step 1:   Substitute all known values into the equation $V = \pi r^2 h$.

$r = 2 \quad \pi = 3.14 \quad h = 5$

$V = \pi r^2 h = 3.14 \, (2)^2 \, (5)$

Step 2:   Solve the equation.

$V = 3.14 \, (2)^2 \, (5) = 3.14 \times 4 \times 5 = 62.8 \ in^3$

Find the volume of the following shapes. Use $\pi = 3.14$.

1.

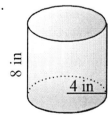

8 in

4 in

2.

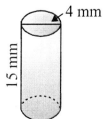

15 mm

4 mm

3.

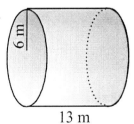

6 m

13 m

4.

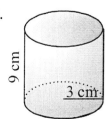

9 cm

3 cm

5.

6 in

20 in

6.

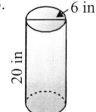

5 cm

12 cm

17.5   Surface Area

The **surface area of a solid** is the total area of all the sides of a solid.

17.6   Cube

There are six sides on a cube. To find the surface area of a cube, find the area of one side and multiply by 6.

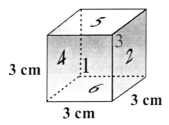

Area of each side of the cube: $3 \times 3 = 9$ cm^2
Total surface area: $9 \times 6 = 54$ cm^2

17.7   Rectangular Prisms

There are 6 sides on a rectangular prism. To find the surface area, add the areas of the six rectangular sides.

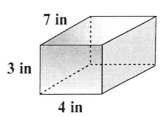

Top and Bottom

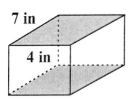

Area of top side:
7 in $\times$ 4 in $= 28$ in^2
Area of top and bottom:
28 in^2 $\times 2 = 56$ in^2

Front and Back

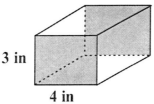

Area of front:
3 in $\times$ 4 in $= 12$ in^2
Area of front and back:
12 in^2 $\times 2 = 24$ in^2

Left and Right

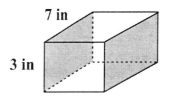

Area of left side:
3 in $\times$ 7 in $= 21$ in^2
Area of left and right:
21 in^2 $\times 2 = 42$ in^2

Total surface area: 56 in^2 + 24 in^2 + 42 in^2 = 122 in^2

Find the surface area of the following cubes and prisms.

1.

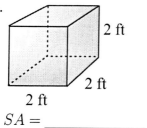

2 ft
2 ft
2 ft

$SA =$ _____

2.

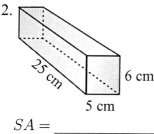

25 cm
6 cm
5 cm

$SA =$ _____

3.

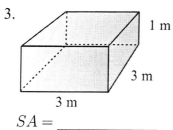

1 m
3 m
3 m

$SA =$ _____

4.

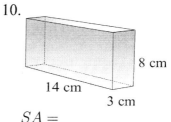

7 mm
7 mm
7 mm

$SA =$ _____

5.

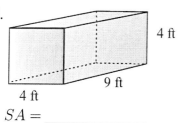

4 ft
9 ft
4 ft

$SA =$ _____

6.

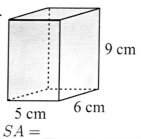

9 cm
5 cm
6 cm

$SA =$ _____

7.

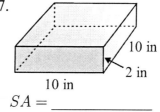

10 in
2 in
10 in

$SA =$ _____

8.

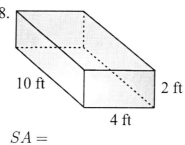

10 ft
2 ft
4 ft

$SA =$ _____

9.

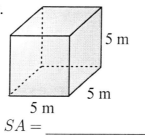

5 m
5 m
5 m

$SA =$ _____

10.

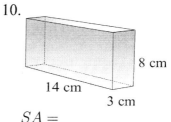

8 cm
14 cm
3 cm

$SA =$ _____

17.8   Cylinders

If the side of a cylinder was slit from top to bottom and laid flat, its shape would be a rectangle. The length of the rectangle is the same as the circumference of the circle that is the base of the cylinder. The width of the rectangle is the height of the cylinder.

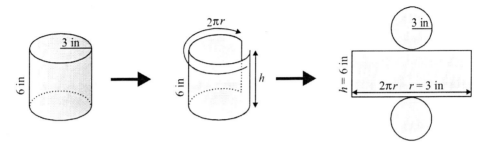

Total Surface Area of a Cylinder $= 2\pi r^2 + 2\pi rh$

Area of top and bottom:
Area of a circle $= \pi r^2$
Area of top $= 3.14 \times 3^2 = 28.26$ in^2
Area of top and bottom $= 2 \times 28.26 = 56.52$ in^2

Area of side:
Area of rectangle $= l \times h$
$l = 2\pi r = 2 \times 3.14 \times 3 = 18.84$ in
Area of rectangle $= 18.84 \times 6 = 113.04$ in^2

Total surface area $= 56.52 + 113.04 = 169.56$ in^2

Find the total surface area of the following cylinders. Use $\pi = 3.14$.

1.

2.
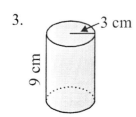

3.
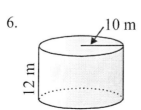

4.
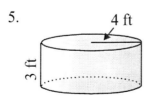

5.
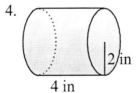

6.
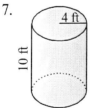

7.
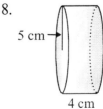

8.
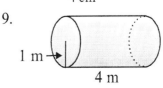

9.

17.9   Solid Geometry Word Problems

1. Robert is using a cylindrical barrel filled with water to flatten the sod in his yard. The circular ends have a radius of 1 foot. The barrel is 3 feet wide. How much water will the barrel hold? The formula for volume of a cylinder is $V = \pi r^2 h$. Use $\pi = 3.14$.

2. Kelly has a rectangular fish aquarium that measures 24 inches wide, 12 inches deep, and 18 inches tall. What is the maximum amount of water that the aquarium will hold?

3. Jenny has a rectangular box that she wants to cover in decorative contact paper. The box is 10 cm long, 5 cm wide, and 5 cm high. How much paper will she need to cover all 6 sides?

4. Gasco needs to construct a cylindrical, steel gas tank that measures 6 feet in diameter and is 8 feet long. How many square feet of steel will be needed to construct the tank? Use the following formulas as needed: $A = l \times w$, $A = \pi r^2$, $C = 2\pi r$. Use $\pi = 3.14$.

5. Jeff built a wooden, cubic toy box for his son. Each side of the box measures 2 feet. How many square feet of wood did he use to build the toy box? How many cubic feet of toys will the box hold?

6. If a cylinder has a radius of 10 cm and a height of 20 cm, what is the volume of the cylinder? The formula for volume of a cylinder is $V = \pi r^2 h$. Use $\pi = 3.14$.

7. Calvin has a rectangular box measuring 6 inches by 8 inches by 14 inches. What is the volume of the box? The formula for volume is $V = lwh$.

8. What is the volume of a cube that is 4 inches tall? The formula for volume of a cube is $V = bwh$.

Chapter 17 Review

Find the volume and/or surface area of the following solids.

1.

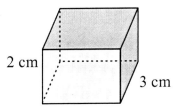

2 cm

3 cm

3 cm

$V =$ _____
$SA =$ _____

2.

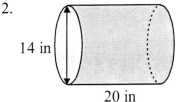

14 in

20 in

Use $\pi = \frac{22}{7}$.
$V =$ _____
$SA =$ _____

3. The sandbox at the local elementary school is 60 inches wide and 100 inches long. The sand in the box is 6 inches deep. How many cubic inches of sand are in the sandbox?

4. A grain silo is in the shape of a cylinder. If the silo has an inside diameter of 10 feet and a height of 35 feet, what is the maximum volume inside the silo? Use $\pi = \frac{22}{7}$.

5. A closed cardboard box is 30 centimeters long, 10 centimeters wide, and 20 centimeters high. What is the total surface area of the box?

6. Siena wants to build a wooden toy box with a lid. The dimensions of the toy box are 3 feet long, 4 feet wide, and 2 feet tall. How many square feet of wood will she need to construct the box?

7. Find the volume of the figure below. Each side of each cube measures 4 feet.

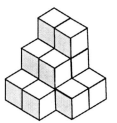

8.

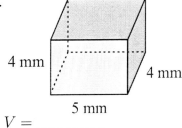

10 cm

18 cm

Use $\pi = 3.14$.
$V =$ _____
$SA =$ _____

9.

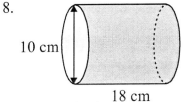

4 mm

4 mm

5 mm

$V =$ _____
$SA =$ _____

10. If a cylinder has a radius of 8 cm and a height of 13 cm, what is the surface area of the cylinder? The formula for surface area of a cylinder is $SA = 2\pi r^2 + 2\pi rh$. Use $\pi = 3.14$.

Chapter 17 Test

1. What is the volume of the following oil tank? Round your answer to the nearest hundredth. Use $\pi = 3.14$.

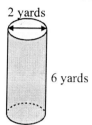

2 yards

6 yards

   A 18.84 yd³
   B 37.68 yd³
   C 44.48 yd³
   D 75.36 yd³

2. What is volume of the box shown below?

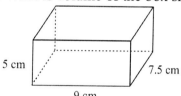

5 cm   7.5 cm

9 cm

   A 22.5 cm³
   B 337.5 cm³
   C 1875 cm³
   D 2250 cm³

3. Find the volume of the cube.

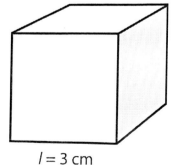

$l = 3$ cm

   A 36 cm³
   B 9 cm³
   C 27 cm³
   D 12 cm³

4. Find the volume of the figure below.

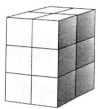

   A 6 units³
   B 12 units³
   C 36 units³
   D 72 units³

5. What is the surface area of a cube whose sides measure 8 cm?

   A 512 cm²
   B 64 cm²
   C 384 cm²
   D 448 cm²

6. If the volume of a cube is 2197 mm³, what is its total surface area?

   A 864 mm²
   B 1183 mm²
   C 169 mm²
   D 1014 mm²

7. What is the surface area of a cylinder whose radius is 6 m and its height is 12 m?

   A 180π m²
   B 144π m²
   C 216π m²
   D 252π m²

8. If the volume of a cylinder is 81π in³ and has a height of 9 in, what is its total surface area?

   A 108π in²
   B 72π in²
   C 120π in²
   D 36π in²

Practice Test 1

1. Simplify: $3^2 + 6(5 - 2) =$

   A 24
   B 27
   C 33
   D 44

   I-1

2. Simplify: $(5 - 2)^2 - 4 - 2 \cdot 6$

   A 5
   B 18
   C -7
   D 90

   I-1

3. Simplify: $4^2 + 8 - 3(8 - 2) + 11$

   A 237
   B 137
   C 17
   D 42

   I-1

4. Simplify: $20 \div 2 - 3^2 - (-2)^2$

   A 24
   B 16
   C 5
   D -3

   I-1

5. Add: $x^2 + 3x + 8$ and $3x^2 + 9$

   A $x^3 + 3x^2 + 3x + 17$
   B $4x^2 + 3x + 17$
   C $7x^2 + 17$
   D $x^2 + 6x + 17$

   I-2

6. Simplify: $(4y^4 + 2y^2 + 7) + (2y^3 + 5y^2 - 4)$

   A $4y^4 + 2y^3 + 7y^2 + 3$
   B $4y^4 + 4y^3 + 5y^2 + 3$
   C $8y^7 + 10y^4 - 28$
   D $8y^{12} + 10y^4 + 3$

   I-2

7. Simplify: $(-3a^2 + 8a - 2) - (-4a^2 - 2a + 6)$

   A $a^2 + 10a - 8$
   B $-7a^2 + 6a + 4$
   C $12a^4 - 16a^2 - 12$
   D $a^2 + 6a - 8$

   I-2

8. Simplify: $(4y^3 - 8y^2 - 5y) - (2y^3 + 5y - 6)$

   A $2y^3 - 8y^2 - 6$
   B $2y^3 - 8y^2 - 10y + 6$
   C $6y^3 - 3y^2 - 10y - 6$
   D $2y^3 - 8y^2 + 6$

   I-2

9. Multiply: $(-3ab^3)(-5a^2b)$

   A $-15a^3b$
   B $15a^3b^4$
   C $15a^2b^3$
   D $15a^3b^3$

   I-3

10. Multiply: $6w(-5w^3 + 2w^2 - 4w)$

   A $-30w^3 + 12w^2 - 24w$
   B $-5w^3 + 2w^2 + 2w$
   C $-30w^4 + 12w^3 - 24w^2$
   D $-30w^4 + 12w^3 + 24w^2$

   I-3

11. Multiply: $(2x^2y)(3xy^3)(-4x^3y^2)$

  A $-24x^6y^6$
  B $-24x^8y^9$
  C $-24x^{27}y^{16}$
  D $24x^6y^7$

I-3

12. Simplify: $(x-2)(3x+1)$

  A $3x^2 - 5x - 2$
  B $3x^2 - 2$
  C $3x^2 + x - 2$
  D $3x^2 - 6x - 2$

I-3

13. Solve for a: $-4a - 12 = -36$

  A 6
  B -6
  C 12
  D -12

II-1

14. Factor: $y^2 - 81$

  A $(y+9)(y+9)$
  B $(y-9)(y-9)$
  C $(y+9)(y-9)$
  D $(y+3)(y-3)$

I-4

15. Factor: $6x^2 - 13x - 5$

  A $(6x-5)(x+1)$
  B $(2x-5)(3x+1)$
  C $(3x-5)(2x+1)$
  D $(x-5)(6x+1)$

I-4

16. Factor: $16x^4 - 1$

  A $(4x^2-1)(4x^2-1)$
  B $(4x^2+1)(4x^2+1)$
  C $(2x-1)(2x-1)(2x+1)(2x+1)$
  D $(4x^2+1)(2x-1)(2x+1)$

I-4

17. Solve: $\dfrac{3x+2}{2} = \dfrac{3x-12}{6}$

  A $x = 1$
  B $x = -1$
  C $x = -3$
  D $x = -6$

II-1

18. Solve the equation $2 - 12a + 5a = a + 6$.

  A $a = -\frac{1}{3}$

  B $a = -\frac{1}{2}$

  C $a = -\frac{1}{4}$

  D $a = -\frac{3}{4}$

II-1

19. Solve for x: $2x + 10 + 4(2x - 1) = -14$

  A $x = -2$
  B $x = -1$
  C $x = -1\frac{4}{5}$
  D $x = -1\frac{2}{10}$

II-1

20. Find the greatest common factor in the following terms:
$11x^5y + 5x^4y^2 - 2x^3y^3 + 6x^2 - 3xy^5$.

  A y
  B x
  C x^2y
  D xy^3

I-4

21. Solve $x^2 + 7x + 12 = 0$ using the quadratic formula.

  A $x = -4, -3$
  B $x = 2, 5$
  C $x = -3, 10$
  D $x = -1, 8$

II-2

22. Solve $3x^2 + 12x + 12 = 0$ using the quadratic formula.

A $x = -2$
B $x = -3, 4$
C $x = 5, 7$
D $x = 6$

II-2

23. Solve $x^2 + 3x - 28 = 0$ by factoring.

A $x = -7, 4$
B $x = 1, 2,$
C $x = -5, 8$
D $x = 0, 3$

II-2

24. Solve: $x^2 + 6x - 7 = 0$

A $x = -7, 1$
B $x = -1, 7$
C $x = 1, 7$
D $x = -1, -7$

II-2

25. What is the solution to the following system of equations?
$y = 3x + 11$
$y = 2x$

A $(-1, -2)$
B $(1, 2)$
C $(-11, -22)$
D $(11, 22)$

II-3

26. What is the solution to the following system of equations?
$5x + 2y = 1$
$2x + 4y = 10$

A $(1, -2)$
B $(3, 1)$
C $(1, 2)$
D $(-1, 3)$

II-3

27. What is the solution to the following system of equations?
$2x - y = 2$
$4x - 9y = -3.$

A $(1.5, 1)$
B $(2, 2)$
C $(-3, -1)$
D $\left(0, \frac{1}{3}\right)$

II-3

28. Find the point of intersection of the two equations.
$4x + 5y = \frac{2}{3}$
$7x - 3y = 9$

A $\left(\frac{2}{3}, -1\right)$
B $\left(-2, \frac{3}{2}\right)$
C $(-1, 0)$
D $\left(1, -\frac{2}{3}\right)$

II-3

29. Solve: $\dfrac{3x + 6}{-2} > -12$

A $x < 24$
B $x > 0$
C $x > 6$
D $x < 6$

II-4

30. Solve for x: $7(2x + 6) - 4(9x + 6) < -26$

A $x > -2$
B $x > 2$
C $x < -2$
D $x < -1$

II-4

31. Solve the inequality $\frac{3}{7}x - 4 < 8$

   A $x < 28$
   B $x < 4$
   C $x > -14$
   D $x > 7$

II-4

32. Is $y = -x + 2$ a function?

   A Yes, because it passes vertical line test.
   B Yes, because it fails horizontal line test.
   C No, because it passes horizontal line test.
   D No, because it fails the vertical line test.

III-1

33. In a science experiment, students hung a cup from a spring and measured the length of the spring when candies were added to it. Their data are shown in the table below. Which statement is true?

# of Candies	Length of Spring
0	0.0
1	1.3
2	2.5
3	5.1
4	6.7
5	8.9
6	10.8
7	12.7
8	14.0
9	15.6

   A The relation is a function because for each x value, there is exactly one y value.
   B The relation is a function because the range values increase.
   C The relation is not a function because only a line can be a function.
   D The relation is not a function because there are two y values for some x values.

III-1

34. Solve the linear inequality $3x - 16 > 5x + 12$.

   A $x > -12$
   B $x > -14$
   C $x < -7$
   D $x < -14$

II-4

35. Which set of points represents a function?

   A $\{(-3, 7), (2, -1), (0, 4), (2, 1)\}$
   B $\{(4, 3), (9, 1), (4, 7), (6, 10)\}$
   C $\{(7, 5), (9, 1), (9, 0)(0, 9)\}$
   D $\{(2, 8), (8, 9), (0, 3), (4, 1)\}$

III-1

36. Which of these equations represents the data in the table?

x	y
1	10
2	18
3	26
4	34

   A $y = 8x$
   B $y = 8x + 2$
   C $y = 8x + 10$
   D $y = 10x - 2$

III-1

37. $f(x) = x - 5$; find $f(3)$.

   A -2
   B 2
   C 8
   D -8

III-2

38. Which of the following is the range of the relation $3x = y - 2$ for the domain $\{-1, 0, 1, 2, 3\}$?

   A $\{-3, 0, 3, 6, 9\}$
   B $\{1, -\frac{2}{3}, -\frac{1}{3}, 0, \frac{1}{3}\}$
   C $\{-3, -2, -1, 0, 1\}$
   D $\{-1, 2, 5, 8, 11\}$

III-2

39. What is the range of the function?

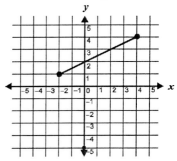

A $-2 \le x \le 4$
B $1 \le x \le 4$
C $-2 \le y \le 4$
D $1 \le y \le 4$

III-2

40. The following graph depicts the height of a projectile as a function of time.

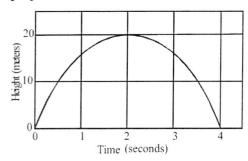

What is the range (R) of this function?

A 0 meters $\le$ R $\le$ 20 meters
B 4 seconds $\le$ R $\le$ 20 meters
C 20 meters $\le$ R $\le$ 4 seconds
D 0 seconds $\le$ R $\le$ 4 seconds

III-2

41. What is the perimeter of a 9-foot by 60-foot rectangular lawn?

A 69 feet
B 138 feet
C 270 feet
D 540 feet

IV-1

42. What is the volume of a wading pool 12 feet long, 6 feet wide, and 6 inches deep?

A 18 cubic feet
B 36 cubic feet
C 216 cubic feet
D 432 cubic feet

IV-1

43. To determine the proper size of air conditioning unit for the classroom represented below, Ms. Waters calculated the volume of the classroom.

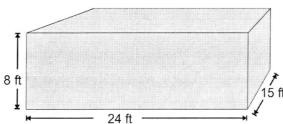

What is the volume of the classroom?

A 47 cubic feet
B 672 cubic feet
C 1334 cubic feet
D 2880 cubic feet

IV-1

44. Find the circumference of the circle below. Use 3.14 for π.

A 18.84 cm
B 37.68 cm
C 75.36 cm
D 113.04 cm

IV-1

45. What is the distance between points $(-2, -4)$ and $(4, -7)$?

A $\sqrt{3}$
B 45
C 3
D $3\sqrt{5}$

IV-2

46. Someone reports a fire at the location plotted on the grid.

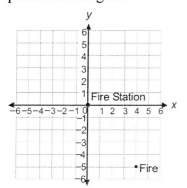

How far is the fire from the fire station?

A 3 miles
B $\sqrt{20}$ miles
C $\sqrt{41}$ miles
D 9 miles

IV-2

47. The coordinates of the endpoints of a line segments are $(3, 1)$ and $(-5, 9)$. Find the coordinates of the midpoint of the line segment.

A $(4, 5)$
B $(-1, 5)$
C $(-2, 4)$
D $(-2, 8)$

IV-2

48. What is the slope of the graph below?

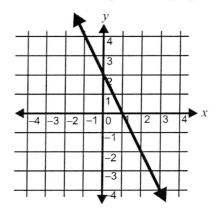

A $-\frac{1}{2}$
B -2
C $\frac{1}{2}$
D 2

IV-2

49. Which equation matches the following graph?

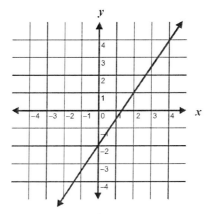

A $y = 3x - 2$
B $y + 2 = -3x$
C $3(y + 2) = 2x$
D $2(y + 2) = 3x$

V-1,4

50. Which of the following is the graph of the equation $y = x + 2$?

A

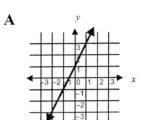

B

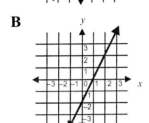

C

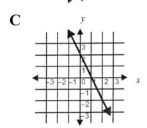

D

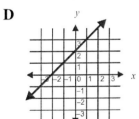

V-1,4

51. Which of the following is a graph of the inequality $6x - 2 \le 5x + 5$?

A

B

C

D

V-3

52. Which is the graph of $f(x) = x$?

A

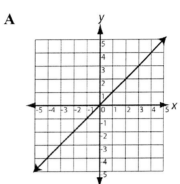

B

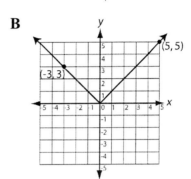

C

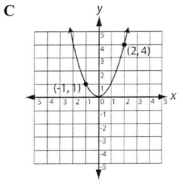

D

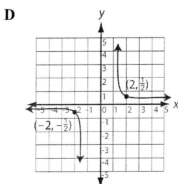

V-1,4

53. The regular price of a stereo (r) is \$560. The stereo is on sale for 25% off. Which equation will help you find the sale price (s) of the stereo?

A $s = r - 0.25$
B $s = r - 0.25s$
C $s = r - 0.25r$
D $s = r - s$

VI-1

54. Celeste earns \$7.00 per hour for the first 40 hours she works this week and time-and-a-half for 5 hours of overtime. Her deductions total \$74.82. Which equation will help Celeste figure her pay?

A $40(7) + 5(7 \times 0.5) - 74.82$
B $40(7 + 5) - 74.52$
C $45(7) - 74.82$
D $7[40 + 5(1.5)] - 74.82$

VI-1

55. Which of these equations represents the graph below?

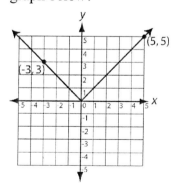

A $y = \sqrt{x}$
B $y = |x|$
C $y = x$
D $y = x^2$

V-1,4

56. Which is the graph of $x - 3y = 6$?

A

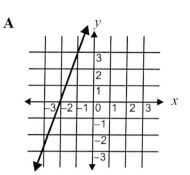

B

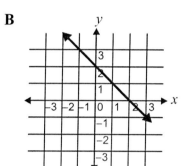

C

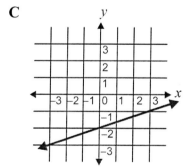

D

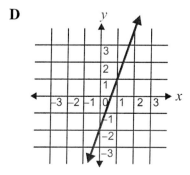

V-1,4

57. Which graph below represents the equation $y = 3x$?

A

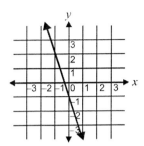

B

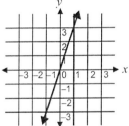

C

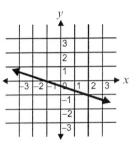

D

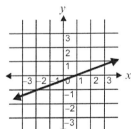

V-1,4

58. What is the equation of the line that includes the point $(3, -1)$ and has a slope of 2?

A $y = -2x - 7$
B $y = -2x - 2$
C $y = -2x + 7$
D $y = 2x - 7$

VI-1

59. Which is the graph of line that passes through the points $(4, 7)$ and $(2, 3)$?

A

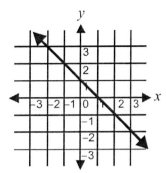

B

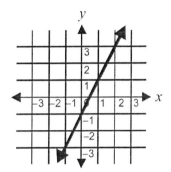

C

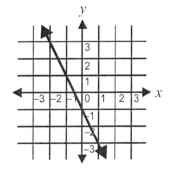

D

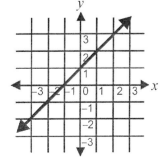

V-2

239

60. Which graph represents a line with a y-intercept of 2 and a slope of $\frac{3}{2}$?

61. Which graph represents a line that passes through the point $(2, 3)$ and a slope of 2?

A

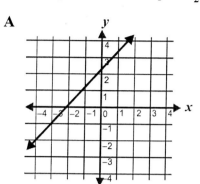

A

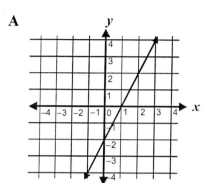

B

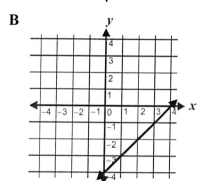

B

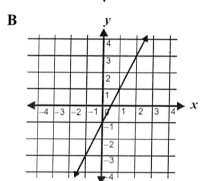

C

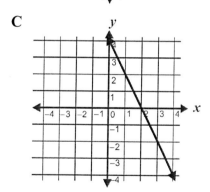

C

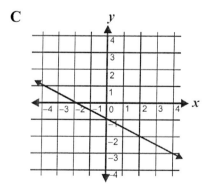

D

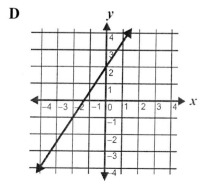

D

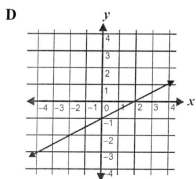

V-2

V-2

62. Which of the following graphs shows a line with a slope $\frac{1}{2}$ that passes through the point $(-1, 1)$?

A

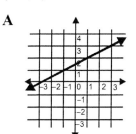

B

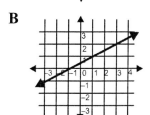

C

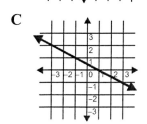

D

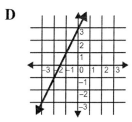

V-2

63. Which of these graphs represents $x < -4$ or $x \geq 1$?

A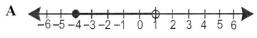

B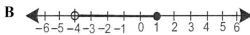

C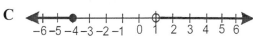

D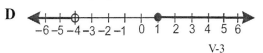

V-3

64. Graph the solution set of $-3 + x \leq -4$.

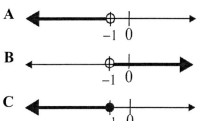

V-3

65. Which of these graphs represents the solution of $1 \leq x + 4 \leq 6$?

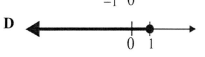

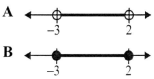

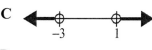

V-3

66. Thomas plans to rent a mower one day next week. He will mow vacant lots and will be paid $20 for each lot he mows. The rental price for the mower is $75 for the day. If l equals the number of lots Thomas mows, which of the following equations could be used to determine the net amount (d) in dollars Thomas will make that day?

A $d - 20l = 75$

B $d = 20l + 75$

C $d - 75 = 20l$

D $d = 20l - 75$

VI-1

241

67. Which of the following equations is represented by this graph?

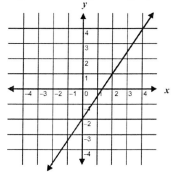

A $y = 2x - 2$

B $y = \frac{3}{2}x - 2$

C $y = \frac{3}{2}x + 2$

D $y = \frac{2}{3}x - 2$

VI-1

68. Find the equation of the line that passes through the points $(2, 3)$ and $(-2, 1)$.

A $y = -\frac{1}{2}x + 4$
B $y = 2x - 1$
C $y = \frac{1}{2}x + 2$
D $y = -2x + 7$

VI-1

69. What is the measure of $\angle WTN$?

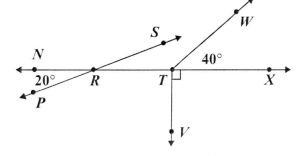

A $40°$
B $50°$
C $140°$
D $180°$

VII-1

70. Angle A and Angle B are supplementary angles. The measure of Angle A is $2x + 15$ and the measure of Angle B is $3x - 25$. What is the measure of Angle A in degrees?

A $19°$
B $38°$
C $89°$
D $91°$

VII-1

71. Name the angle that is vertical to $\angle EAF$.

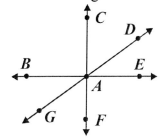

A $\angle EAD$
B $\angle CAB$
C $\angle BAG$
D $\angle GAF$

VII-1

72. On the diagram below, which angles are corresponding angles?

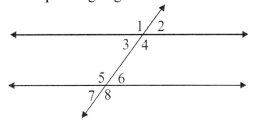

A 2 and 5
B 4 and 8
C 3 and 5
D 6 and 7

VII-1

73.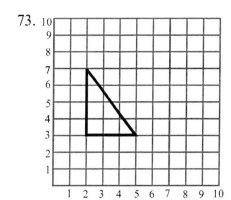

Carlo has drawn a right triangle on the grid above. What is the perimeter of the triangle?

A 12 units
B 14 units
C 18 units
D 20 units

VII-2

74. Olivia drives 10 miles south and then 8 miles east to get to work. If there were a road that went in a straight line from her house to work, how many miles would she save getting to work in the morning?

A 12.8 miles
B 9 miles
C 5.2 miles
D 18 miles

VII-2

75. The following triangles are similar. What is the measure of side x?

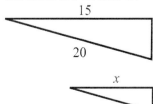

A 8
B 9
C 10
D 16

VII-3

76. A utility pole is to be fitted with a guy wire. One end of the wire will be attached at ground level, 8 feet from the base of the pole. The other end will be attached to the pole, 15 feet above ground level. What is the length of the wire?

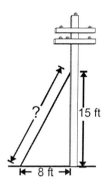

A 17 feet
B 18 feet
C 19 feet
D 20 feet

VII-2

77. Using the Pythagorean Theorem, what is the measure of side c of the triangle below? Round to the nearest tenth?

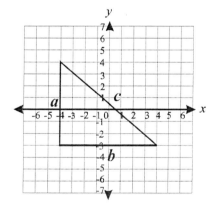

A 15
B 7.5
C 8
D 10.6

VII-2

78. △ABC is similar to △DEF. What is the value of x?

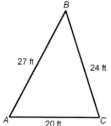

A 12 feet
B 13 feet
C 14 feet
D 15 feet

VII-3

79. The measure of the sides of triangles are given below. Which of the following triangles is similar to a triangle with sides 2.4, 7, and 7.4?

A 3, 8.75, 9
B 4, 8.4, 9.2
C 6, 17.5, 18.5
D 12, 14, 14.8

VII-3

80. The following two figures are similar.

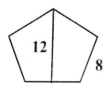
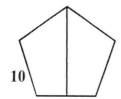

What is the height of the second pentagon?

A 15
B 14
C 9.6
D 20

VII-3

81. An advertising sign is an equilateral triangle with sides 10 feet long.

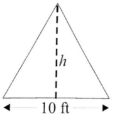

What is the approximate height of the sign?

A 7 feet, 10 inches
B 8 feet, 10 inches
C 8 feet, 8 inches
D 9 feet, 1 inches

VII-4

82. A Ferris wheel has a radius of 14 feet. How far will you travel if you take a ride that goes around six times? Use $\pi = \frac{22}{7}$.

A 528 feet
B 616 feet
C 3,696 feet
D 12,936 feet

VII-4

83. Sunshine Feed and Grain Supply has a huge cylindrical container that holds approximately 4,000 cubic feet of corn. The diameter of the container is 10 feet. What is the approximate height of the container?

A 13 feet
B 40 feet
C 50 feet
D 160 feet

VII-4

84. Which set of numbers has the greatest range?

A {95, 86, 78, 62}
B {90, 65, 83, 59}
C {32, 29, 44, 56}
D {29, 35, 49, 51}

VII-5

244

85. What is the volume of the figure below?

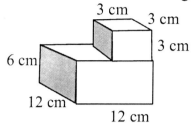

A 21 cm³
B 39 cm³
C 216 cm³
D 891 cm³

VII-4

86. What is the median of 27, 32, 16, 31, 9, and 37?

A 31
B 16
C 28
D 29

VII-5

87. George has scores of 76, 78, 79, and 67 on four history tests. What is the lowest score George can have on the fifth test to have an average score of 80?

A 85
B 90
C 95
D 100

VII-5

88. Lisa's exam scores for history are listed below. What is her average score for the tests?

Test 1	95
Test 2	105
Test 3	80

A 91
B 92
C 100
D 93

VII-5

89. Tim tosses three nickels on the ground. What is the probability that all three will show "heads"?

A $\frac{1}{8}$
B $\frac{3}{8}$
C $\frac{1}{2}$
D $\frac{8}{27}$

VII-6

90. A box contains spools of thread: 3 spools of red, 4 spools of blue, 2 spools of green, and 3 spools of yellow. What is the probability of reaching in the box without looking and picking a red spool?

A $\frac{1}{4}$
B $\frac{1}{5}$
C $\frac{2}{7}$
D $\frac{3}{4}$

VII-6

91. Farmer MacDonald's hen lays eggs at the rate of approximately 50 eggs in three months. If the hen continues at this rate, approximately how many eggs will she lay in 4 years?

A 67
B 200
C 600
D 800

VII-7

92. The length of a rectangle is 3 more than twice the width. The perimeter is 36. What is the length?

A 5
B 10
C 13
D 15

VII-8

93. If Steven rolls two dice, what is the probability that the dice total 7 or 11?

A $\frac{2}{9}$

B $\frac{2}{12}$

C $\frac{1}{6}$

D $\frac{1}{3}$

VII-6

94. Tina has a bag of jelly beans of four flavors. There are 6 licorice, 4 grape, 3 sour apple, and 5 mint jelly beans in the bag. Tina prefers the fruit-flavored jelly beans. If she selects two jelly beans at random from the bag one at a time, what is the probability that the two jelly beans will be grape?

A $\frac{2}{51}$

B $\frac{61}{153}$

C $\frac{4}{9}$

D $\frac{4}{81}$

VII-6

95. Brice can type 22 pages in 3 hours. At this rate approximately how long would it take to type one page?

A 7.3 minutes

B 8.2 minutes

C 12.2 minutes

D 13.6 minutes

VII-7

96. To make a disinfecting solution, Alana mixed 2 cups of bleach with 5 cups of water. What is the ratio of bleach to the total amount of disinfecting solution?

A 2 to 3

B 2 to 5

C 2 to 7

D 2 to 10

VII-7

97. For every 4 fish that Alice has in her pond, she must have five plants for them. If she only has 75 plants, what is the total number of fish she can have?

A 60

B 94

C 135

D 169

VII-7

98. The sum of two numbers is fourteen. The sum of six times the smaller number and two equals four less than the product of three and the larger number. Find the two numbers.

A 6 and 8

B 5 and 9

C 3 and 11

D 4 and 10

VII-8

99. There were three brothers. Fernando was two years older than Pedro. Pedro was two years older than Samuel. Together their ages add up to 63 years. How old is Samuel?

A 17

B 19

C 21

D 23

VII-8

100. Ralph has borrowed $600 for 2 years at an annual rate of 5%.
Use the formula $I = PRT$ to find the amount of interest he will pay.

A $15

B $24

C $60

D $150

VII-8

Practice Test 2

1. Which order of operations should be used to simplify the following expression:

$6(7 - 2) \div 2 + 5$

  A subtract, multiply, divide, add
  B subtract, add, multiply, divide
  C add, subtract, multiply, divide
  D multiply, subtract, divide, add

I-1

2. Simplify: $25 - 4^2 - (3 - 8)$

  A -2
  B 14
  C 26
  D 16

I-1

3. Simplify: $24 \div 3 - 2^2$

  A 4
  B 36
  C 24
  D 12

I-1

4. Simplify $4^2 \div 8 \times 2 - 4$

  A -3
  B -4
  C -1
  D 0

I-1

5. Add $x^4 + 2x^2 + 6$ and $4x^3 + 5x^2 - 3$

  A $x^4 + 6x^3 + 5x^2 + 3$
  B $5x^3 + 7x^2 + 3$
  C $x^4 + 4x^3 + 6x^2 - 3$
  D $x^4 + 4x^3 + 7x^2 + 3$

I-2

6. Find: $(3y^3 + 5y^2 - 8) + (4y^3 - 6y^2 + 3)$

  A $7y^3 - y^2 - 5$
  B $7y^3 - 6y^2 - 5$
  C $3y^3 + 13y^2 - 5$
  D $3y^3 - y^2 - 5$

I-2

7. Find: $(2x^3 + 4x^2 + 7x) - (3x^3 - 2x - 5)$

  A $-x^3 + 4x^2 + 5x - 5$
  B $5x^3 + 4x^2 + 5x - 5$
  C $-x^3 + 4x^2 + 9x + 5$
  D $5x^3 + 4x^2 + 9x + 5$

I-2

8. Find: $(-7b^4 - 5b^3 + 8b - 4) + (-2b^4 - 4b^3 + 10)$

  A $-9b^4 - 9b^6 + 8b + 6$
  B $-9b^8 - 9b^6 + 8b + 14$
  C $-5b^8 - 9b^6 + 8b - 6$
  D $-9b^4 - 9b^3 + 8b + 6$

I-2

9. Simplify: $-3(x - 5)^2$

  A $-3x^2 - 10x + 25$
  B $x^2 - 10x + 25$
  C $3x^2 - 30x + 75$
  D $-3x^2 + 30x - 75$

I-3

10. Multiply: $(-4wx)(-2w^4x^3)$

  A $-8w^5x^4$
  B $8w^5x^4$
  C $-8w^4x^3$
  D $8w^4x^4$

I-3

11. Multiply: $4y(-y^2 - 3y + 2)$

   A $-4y^3 - 12y^2 + 8y$
   B $-4y^2 - 12 + 8$
   C $-4y^3 - 3y^2 + 2y$
   D $4y^3 + 12y^2 + 8y$

   I-3

12. Multiply: $-6a^3(-2ab^2 + 5a^2b - 6a^3)$

   A $12a^3b^2 - 30a^6 + 36a^9$
   B $12a^4b^2 - 30a^5b + 36a^6$
   C $-12a^3b^2 + 30a^6 - 36a^9$
   D $-12a^4b^2 + 30a^5b - 36a^6$

   I-3

13. Factor: $b^2 - 2b - 8$

   A $(b - 4)(b + 4)$
   B $(b - 2)(b + 4)$
   C $(b + 2)(b - 4)$
   D $(b - 2)(b - 2)$

   I-4

14. Factor: $4x^2 + x - 3$

   A $(2x + 3)(2x - 1)$
   B $(4x - 3)(x + 1)$
   C $(2x - 3)(2x + 1)$
   D $(4x + 3)(x - 1)$

   I-4

15. Which expression is a factor of $(15x^2 + 11x - 14)$?

   A $x - 2$
   B $5x + 7$
   C $3x + 2$
   D $15x - 1$

   I-4

16. Factor $18x^2 - 32$.

   A $2(3x - 4)(3x + 4)$
   B $(6x - 2)(3x + 2)$
   C $(9x + 3)(2x + 1)$
   D $4(4x^2 - 8)$

   I-4

17. What is the value of x if $4x - 2 = 22$?

   A 5
   B 20
   C 16
   D 6

   II-1

18. Solve for x: $4x + 6 = -26$

   A -5
   B -6
   C -8
   D 9

   II-1

19. Find b: $b - 7(b + 2) = 10$

   A $-\dfrac{4}{3}$
   B -4
   C 4
   D -8

   II-1

20. Solve: $\dfrac{x + 2}{7} = \dfrac{x + 4}{21}$.

   A $x = -1$
   B $x = 2$
   C $x = -4$
   D $x = 7$

   II-1

21. Solve: $x^2 + 12x + 27 = 0$

   A $x = -9, -3$
   B $x = -21, 9$
   C $x = -3, 9$
   D $x = -21, 3$

   II-2

22. Solve $x^2 - 4x - 12 = 0$ using the quadratic formula.

   A $x = -2, 2$
   B $x = -12, 1$
   C $x = -2, 6$
   D $x = -5, 7$

II-2

23. Solve $3x^2 + 5x - 2 = 0$ by factoring.

   A $x = \frac{1}{3}, 2$

   B $x = -2, \frac{1}{3}$

   C $x = -\frac{1}{3}, 2$

   D $x = -2, -\frac{1}{3}$

II-2

24. Solve: $x^2 + 12x - 36 = 9$

   A $x = -6, 6$
   B $x = -15, 3$
   C $x = -87, 75$
   D $x = 2, 4$

II-2

25. What is the solution to the two graphs $y = 2x + 2$ and $y = -5x - 12$ shown below?

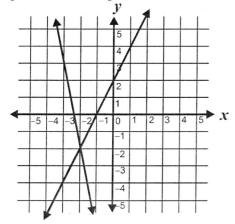

   A $(-2, 2)$
   B $(-2, -2)$
   C $(2, 2)$
   D $(2, -2)$

II-3

26. Which ordered pair is a solution for the following system of equations?

$$-3x + 7y = 25$$
$$3x + 3y = -15$$

   A $(-13, -2)$
   B $(-6, 1)$
   C $(-3, -2)$
   D $(-20, -5)$

II-3

27. What is the solution to the equations $x - y = 1$ and $x + y = 6$?

   A $\left(\frac{7}{4}, \frac{3}{4}\right)$

   B $\left(\frac{10}{3}, \frac{8}{3}\right)$

   C $\left(\frac{5}{2}, \frac{7}{2}\right)$

   D $\left(\frac{7}{2}, \frac{5}{2}\right)$

II-3

28. What is the solution of the following system of linear equations?
$$2y + 6 = x$$
$$y - 3x = -13$$

   A $(8, 1)$
   B $(4, -1)$
   C $(3, -4)$
   D $(12, 3)$

II-3

29. Solve: $-6x + 3 \leq 2x - 13$

   A $x \leq 2$
   B $x \leq 2.5$
   C $x \geq 2$
   D $x \geq 2.5$

II-4

30. Solve: $-6 - x \geq 7$

   A $x \geq -13$
   B $x \leq 13$
   C $x \leq -13$
   D $x \geq 13$

II-4

31. Solve: $10 + 3(2x - 6) \leq 8 - 2x$

    A $x \leq 2$
    B $x \geq -2$
    C $x \leq -2$
    D $x \geq 2$ II-4

32. Solve: $-\frac{4}{5}x \geq 8$

    A $x \geq 10$
    B $x \geq 5$
    C $x \leq -10$
    D $x \geq 10$ II-4

33. Is the graph below a function?

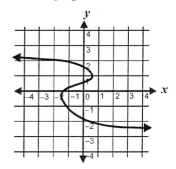

    A Yes, because it passes the vertical line test.
    B Yes, because it passes the horizontal line test.
    C No, because it fails the vertical line test.
    D No, because it fails the horizontal line test.   III-1

34. The following is a table of the cost of mail at specified ounces. Which equation best represents the table?

x (oz.)	1	2	3	4	5
y (\$)	0.80	0.97	1.14	1.31	1.48

    A $y = 0.17x + 0.63$
    B $y = 0.63x + 0.17$
    C $y = 0.17x$
    D $y = 0.63x$ III-1

35. Is the following graph a function?

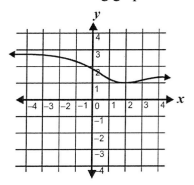

    A Yes, because it passes the vertical line test.
    B Yes, because it passes the horizontal line test.
    C No, because it fails the vertical line test.
    D No, because it fails the horizontal line test.   III-1

36. What is the range of the following function? $\{(0, 1)(1, 4)(2, 6)(3, 9)\}$

    A $\{0, 1, 2, 3\}$
    B $\{1, 4, 6, 9\}$
    C $\{0, 1, 4\}$
    D $\{0, 1, 3, 9\}$ III-2

37. Find the range of the following function for the domain $\{-2, -1, 0, 3\}$.
$$y = \frac{2 + x}{4}$$

    A $\left\{0, \frac{3}{4}, 1, \frac{5}{4}\right\}$

    B $\left\{0, \frac{1}{4}, \frac{1}{2}, \frac{5}{4}\right\}$

    C $\left\{1, -\frac{1}{4}, \frac{1}{2}, \frac{5}{4}\right\}$

    D $\left\{\frac{1}{4}, \frac{3}{4}, \frac{1}{2}, \frac{5}{4}\right\}$ III-2

38. What is the range of the function $y = 2x - 8$ for the domain $\{10, 11, 12, 13\}$?

  A $\{9, 9\frac{1}{2}, 10, 10\frac{1}{2}\}$
  B $\{5, 6, 7, 8\}$
  C $\{9, 10, 11, 12\}$
  D $\{12, 14, 16, 18\}$

III-2

39. Which of these functions describes the mapping below?

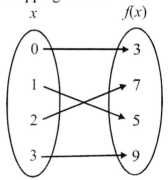

  A $f(x) = 3x + 2$
  B $f(x) = 5x - 2$
  C $f(x) = 2x + 3$
  D $f(x) = x + 3$

III-1

40. Find the area of the trapezoid below.

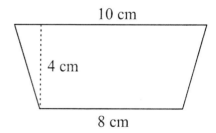

  A 22 square centimeters
  B 36 square centimeters
  C 72 square centimeters
  D 320 square centimeters

IV-1

41. Jack is going to paint the ceiling and four walls of a room that is 10 feet wide, 12 feet long, and 10 feet from floor to ceiling. How many square feet will he paint?

  A 120 square feet
  B 560 square feet
  C 680 square feet
  D 1,200 square feet

IV-1

42. For $f(x) = 3x^2 - 5x$, find $f(-3)$.

  A 12
  B -6
  C 42
  D 3

III-2

43. A preschool is required to have a playground of at least 900 square feet. Which of the following would be satisfactory measurements for a playground for the school?

  A 30 feet by 32 feet
  B 27 feet by 30 feet
  C 15 feet by 40 feet
  D 10 feet by 80 feet

IV-1

44. What is the area of the triangle?

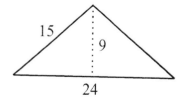

  A 54 square units
  B 67.5 square units
  C 108 square units
  D 216 square units

IV-1

45. What is the slope of a line passing through the points $(3, 6)$ and $(5, 1)$?

A $-\dfrac{5}{2}$

B $-\dfrac{4}{3}$

C $-\dfrac{3}{4}$

D $\dfrac{2}{5}$

IV-2

46. The coordinates for the endpoints of a line segment are $(-5, 13)$ and $(-9, 21)$. Find the coordinates for the midpoint of the line segment.

A $(-7, 17)$
B $(-4, 4)$
C $(-2, 4)$
D $(4, 17)$

IV-2

47. The coordinates of a line segment are $(1, 6)$ and $(11, -4)$. What are the coordinates for the midpoint of the line segment?

A $(6, 1)$
B $(10, 2)$
C $(5, 1)$
D $(12, 2)$

IV-2

48. One endpoint of a line segment $\overline{AB}$ is $B(0, -5)$ and the midpoint of line segment $\overline{AB}$ is $(-2, 3)$. What is the length of line segment $\overline{AB}$?

A $4\sqrt{17}$
B $2\sqrt{17}$
C $8\sqrt{4}$
D $16\sqrt{2}$

IV-2

49. Which is the graph of $f(x) = |x|$?

A

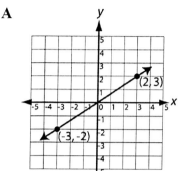

B

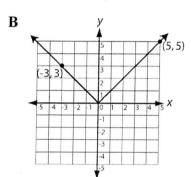

C

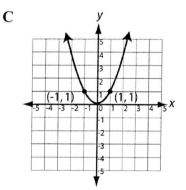

D

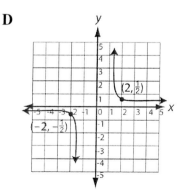

V-1,4

50. Which graph below shows the function $f(x) = -3x - 1$?

A

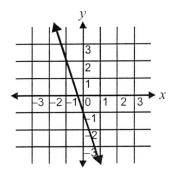

B

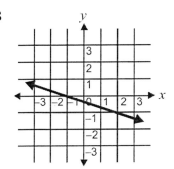

C

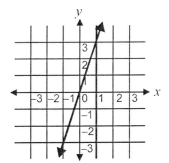

D

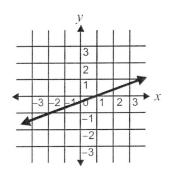

V-1,4

51. Below is the graph of which equation?

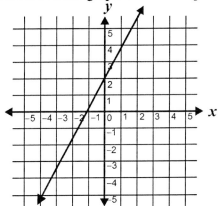

A $y = x + 2$

B $y = 2x - 2$

C $y = 2x + 2$

D $y = x - 2$

V-1,4

52. Which of these equations represents the graph below?

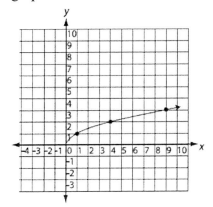

A $y = x$

B $y = x^2$

C $y = |x|$

D $y = \sqrt{x}$

V-1,4

53. Which graph represents the equation $-2y = -2x + 8$?

A

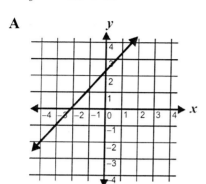

B

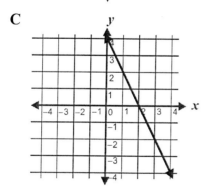

C

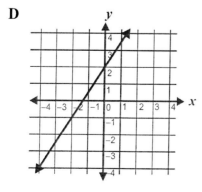

D

54. Which graph represents the equation $-2x + y = -2$?

A

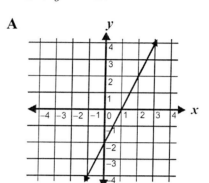

B

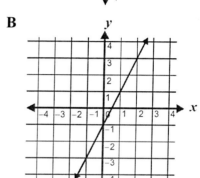

C

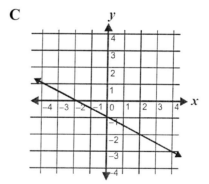

D

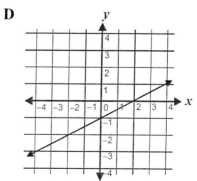

V-1,4

V-1,4

55. Which graph shows a line with a slope of $\frac{5}{2}$, passing through the point $(1, -3)$?

A

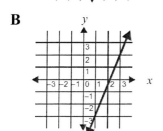

B

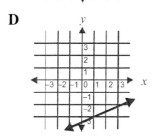

C

D

V-2

56. Which of these graphs represents $x \le -1$?

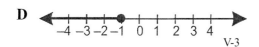

A

B

C

D

V-3

57. Which equation passes through $(3, 0)$ and $(-2, 4)$?

A

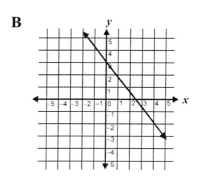

B

C

D

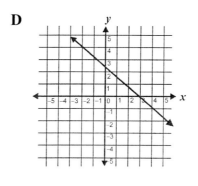

V-2

58. Which graph represents a line containing the points $(-1, -2)$ and $(-4, 3)$?

A

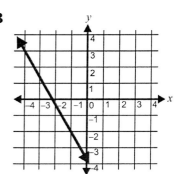

B

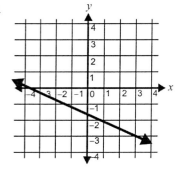

C

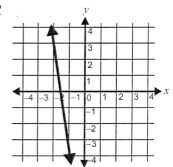

D

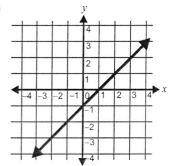

59. Which graph shows a line with a slope of -3 and a y-intercept of 3?

A

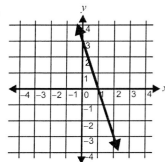

B

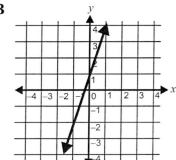

C

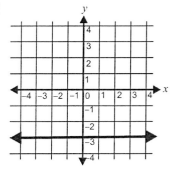

D

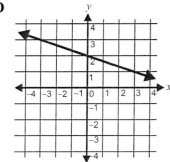

V-2

V-2

60. Which of these graphs represents the solution of $-2 \le x < 3$?

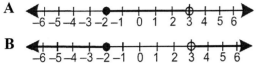

V-3

61. Which of these graphs represents the solution of $-2x \ge 6$?

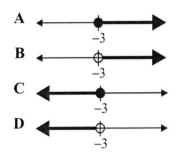

V-3

62. Which of these graphs represents the solution of $3x - 1 < 5$ or $4 - x \le 5$?

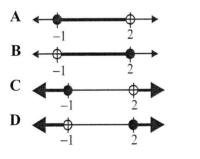

V-3

63. Della is renting a car for the day. The rental fee (y) is \$30 plus \$0.25 per mile (m). Which of the following equations represents this cost?

A $y = 0.30m + 25$
B $y = 30m + 0.25$
C $y = 0.25m + 30$
D $y = m(0.25 + 30)$

VI-1

64. Peter buys T-shirts wholesale by purchasing 125 of them for \$250. To sell the shirts at a profit, Peter sells them with his own designs in sets of five. Which formula will calculate Peter's cost, x, for the shirts in each set?

A $\dfrac{5}{125}x = \$250$

B $\dfrac{\$250}{5} = x$

C $x = 5\left(\dfrac{\$250}{125}\right)$

D $125x = \$250$

VI-1

65. Below is the table of gas prices at specified gallons. Which equation best represents the table below?

x (gal.)	0	2	3	7	15
$y(\$)$	0.00	5.68	8.52	19.88	42.60

A $y = 2.83x + 3$
B $y = 2.84x + 3$
C $y = 2.84x$
D $y = 2.84x - 1$

VI-1

66. Which equation of the line shown in the graph below?

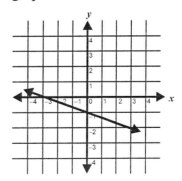

A $x = -3y + 3$
B $x = -\frac{1}{3}y + 1$
C $y = -\frac{1}{3}x - 1$
D $-3x + y - 1 = 0$

VI-1

67. A plumber charges $39 to come to your house and $28 for each hour after that. If you received a bill for $193, write an equation for the number of hours, h, worked.

A $28 + 39h = 193$
B $39 + 28h = 193$
C $193 - 28 = 39h$
D $193 = 39 - 28h$

VI-1

68. Which equation matches the graph of the line below?

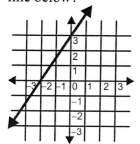

A $y = \frac{3}{2}x - 2$
B $y = \frac{3}{2}x + 3$
C $y = \frac{2}{3}x + 3$
D $y = \frac{2}{3}x - 2$

VI-1

69. What is the value of x? Lines l and m are parallel to each other.

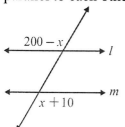

A $75°$
B $95°$
C $105°$
D $150°$

VII-1

70. What is the relationship between $\angle 1$ and $\angle 4$?

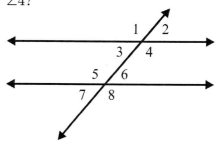

A They are corresponding angles.
B They are vertical angles.
C They are alternate interior angles.
D They are supplementary angles.

VII-1

71. If the measure of $\angle 2$ is $50°$, what is the measure of $\angle 4$?

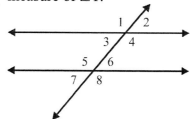

A $50°$
B $100°$
C $130°$
D Not enough information given.

VII-1

72. What is the measure, in degrees, of $\angle CBE$?

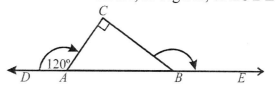

A $30°$
B $60°$
C $120°$
D $150°$

VII-1

73. In the right triangle below, what is the value of x?

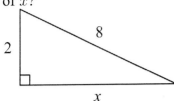

A $2\sqrt{15}$
B 68
C $2\sqrt{17}$
D $\sqrt{10}$

VII-2

74. Find the length of the missing side of the triangle below.

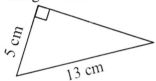

A 10 cm
B 11 cm
C 12 cm
D 15 cm

VII-2

75. Sally made a scale drawing of her garage.

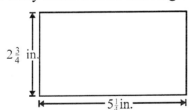

The actual length of the garage is 21 feet. What is the width of the garage?

A 10 feet
B 11 feet
C 12 feet
D 14 feet

VII-3

76. Jill is flying a kite on 100 feet of string. She holds the end of the kite string to the ground while Jack measures the distance to a point directly under the kite.

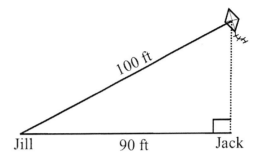

If the distance from Jill to Jack is 90 feet, how high is the kite above the ground? Answer to the nearest whole foot.

A 10 ft
B 19 ft
C 44 ft
D 55 ft

VII-2

77. A fiber-optic cable is to be installed between point m and point n.

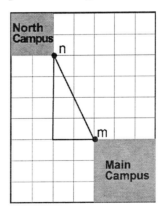

What will be the approximate length of the cable?

A 3.9 km
B 4.5 km
C 4.7 km
D 4.9 km

VII-2

78. Find the missing side from the following similar triangles.

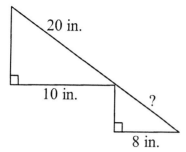

A 8
B 16
C 24
D 32

VII-3

79. The following regular octagons are similar to one another. Find the length from one vertex to its opposite vertex in the smaller octagon.

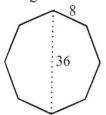

A 13.5
B 28
C 31
D 108

VII-3

80. A box has a volume of 1440 cubic inches, a height of 10 inches, and a square base. What is the length of one side of the base?

A 6 inches
B 12 inches
C 24 inches
D 48 inches

VII-4

81. Bernard wants to determine the height of a monument tower. He has made a pinhole camera from a box and placed it so that an image of the monument is projected on the inside of the box opposite the pinhole. The figure below is not drawn to scale.

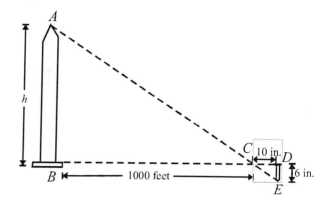

Is $\triangle ABC$ similar to $\triangle EDC$?

A Yes
B No
C Yes, because the two triangles are congruent to each other.
D Cannot be determined

VII-3

82. Darin is playing a dart game at the county fair. At the booth, there is a spinning board completely filled with different colors of balloons. There are 6 green, 4 burgundy, 5 pink, 3 silver, and 8 white balloons. Darin aims at the board with his dart and pops one balloon. What is the probability that the balloon popped is green?

A $\frac{1}{13}$

B $\frac{10}{13}$

C $\frac{3}{13}$

D $\frac{2}{13}$

VII-6

83. What is the area of a triangle with a base of $6x - 4$ and a height of x?

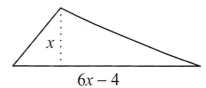

A $3x^2 - x$

B $3x^2 - 1$

C $3x^2 - 2x$

D $3x^2 - 2$

VII-4

84. Crystal has a rectangular sheet of wrapping paper measuring 30 inches along one side. The perimeter of the paper is 132 inches. What is the length of the other side?

A 36 inches

B 38 inches

C 51 inches

D 72 inches

VII-4

85. The living room in Ty's house has 168 square feet of floor space. His family is building an addition to this room that measures 14 feet long and 8 feet wide. What will be the total square feet of the living room with the new addition?

A 112 square feet

B 180 square feet

C 270 square feet

D 280 square feet

VII-4

86. Consider the following set of data.

$\{1, 2, 3, 4, 6, 7, 8, 9, 9, 11, 11, 11, 12\}$

What is the mode of this set of data?

A 7.3

B 10

C 11

D 12

VII-5

87. Justin recorded the weights of 6 wrestlers. Their weights, in kilograms, are given below.

66, 97, 52, 53, 76, 105

What is the median weight of the 6 wrestlers?

A 52.5 kilograms

B 71.0 kilograms

C 85.5 kilograms

D 86.5 kilograms

VII-5

88. Eddie averaged 34 points per game in the 12 regular season games. He averaged 42 points per game in the 3 tournament games. What is Eddie's overall average?

A 44.5

B 38

C 37.2

D 35.6

VII-5

89. The teacher has 3 red pens, 8 black pens, and 4 blue pens all the same size in a box. What is the probability of reaching in without looking and picking out a red pen?

A $\dfrac{1}{15}$

B $\dfrac{1}{3}$

C $\dfrac{1}{4}$

D $\dfrac{1}{5}$

VII-6

90. The monthly incomes of 5 individuals are shown below:

$2,540; $9,985; $2,789; $2,748; $2,065

Which of the following best represents the approximate difference between the mean and the median of these five monthly incomes?

A $1,200
B $1,300
C $1,900
D $2,700

VII-5

91. Molly had two boxes of new pencils, one containing 160 and the other containing 40 pencils. The percent of red pencils in each box was 40% and 5% respectively. She combined the pencils into one box. Molly then selected 1 pencil without looking. What is the probability she chose a red pencil?

A 22.5%
B 33%
C 45%
D 66%

VII-6

92. If Charles spins the spinner pictured below, which of the following is most likely to happen?

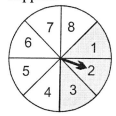

A It will land on an unshaded number.
B It will land on an even number.
C It will land on an odd shaded number.
D It will land on an odd number.

VII-6

93. If 3 out of 4 people use a certain headache medicine, how many in a city of 150,400 will use this medicine?

A 118,200
B 37,600
C 50,133
D 112,800

VII-7

94. The scale of a map is $\frac{1}{2}$ inch = 40 miles. If two towns are 6 inches apart on the map, how many miles apart are they?

A 480 miles
B 120 miles
C 80 miles
D 30 miles

VII-7

95. The conversion of U.S. currency to Peruvian currency is 1 U.S. dollar = 1.2 soles. If you exchange 53.65 U.S. dollars, how many soles will that be?

A 41.26
B 64.38
C 44.71
D 53.65

VII-7

96. Suppose that y varies directly with x and $x = 14$ when $y = 2$. What is y when $x = 3$?

A 1
B $\frac{5}{14}$
C $\frac{14}{3}$
D $\frac{3}{7}$

VII-7

97. There are 5 more than twice as many boys in the weightlifting class as there are girls. There are 38 students in the class altogether. How many girls are in the class?

A   8
B   11
C   12
D   15

VII-8

98. Lloyd wanted to see how many of each candy flavors he had in his lunch box. He counted 25 pieces out of which he had 3 more lemon than grape, and 5 more grape than strawberry. How many strawberry candies did Lloyd have?

A   4
B   9
C   12
D   15

VII-8

99. There are 4 more than 3 times the number of pennies than there are number of dimes. The change totals to $1.21. How many coins are there?

A   7
B   27
C   31
D   40

VII-8

100. Michael borrowed $4,500 from his dad to buy a used car. He agreed to pay his dad back in one year with 6% simple interest. How much interest will Michael pay? Use the formula $I = PRT$.

A $27.00
B $270.00
C $24.00
D $240.00

VII-8

Index